FOSSIL TREASURES OF FOULDEN MAAR

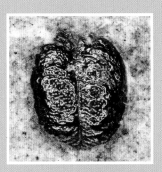

FOSSIL TREASURES OF FOULDEN MAAR

A window into Miocene Zealandia

Daphne Lee, Uwe Kaulfuss and John Conran

OTAGO UNIVERSITY PRESS
Te Whare Tā o Te Wānanga o Ōtākou

PICTURE CREDITS

DCM Dallas Mildenhall

DEL Daphne Lee

EG Liz Girvan

EMK Liz Kennedy

JGC John Conran

JHR Jeffrey Robinson

JKL Jon Lindqvist

JMB Jennifer Bannister

SER Stephen Read

UK Uwe Kaulfuss

WGL Bill Lee

Published by Otago University Press
Te Whare Tā o Te Wānanga o Ōtākou
533 Castle Street
Dunedin, New Zealand
university.press@otago.ac.nz
www.otago.ac.nz/press

First published 2022
Text copyright © Daphne Lee, John Conran and
Uwe Kaulfuss

The moral rights of the authors have been asserted.

ISBN 978-1-99-004835-7

Editor and indexer: Imogen Coxhead
Design/layout: Fiona Moffat

Printed in China through Asia Pacific Offset

MIX
Paper from
responsible sources
FSC® C136333
FSC
www.fsc.org

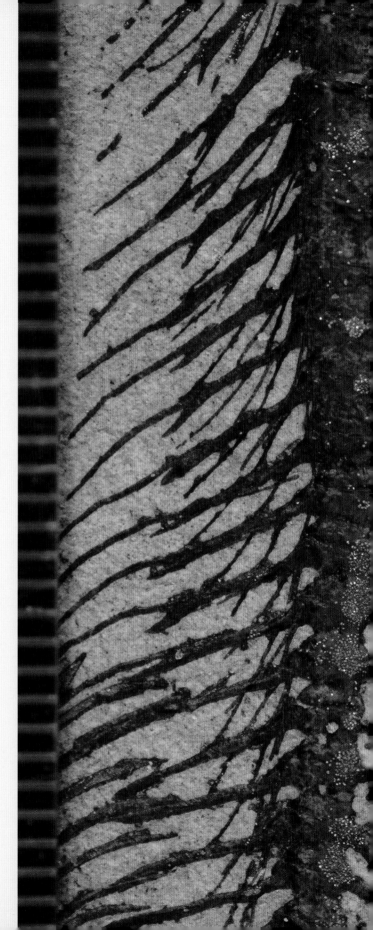

CONTENTS

FOREWORD

In 2009, Uwe Kaulfuss, then a research student at the University of Otago, sent me some pictures of fossil spiders discovered at Foulden Maar. These were the first fossil spiders ever found in New Zealand. Spider remains are very rare in the geological record because a spider's anatomy has no hard parts. The presence of soft-bodied organisms at a fossil locality defines that place as somewhere very special: a *Konservat-Lagerstätte*. I was excited!

Being fragile, the specimens could not be mailed, but Uwe sent more photographs and eventually we published an article together. Some of the fossils were identifiable only as 'arachnid' or 'spider'. One, however, was related to the modern spider family Idiopidae, a family that is thought to have dispersed to New Zealand from Australia. Evidence that this family existed in New Zealand during the Miocene suggests a time frame for this event. The presence of spiders and other preserved terrestrial biota at Foulden Maar refutes the hypothesis that the whole of New Zealand was under water at the time.

In 2019, I travelled to New Zealand for an International Arachnological Congress. Daphne Lee arranged for me to see the fossil spiders, and organised a field trip to Foulden Maar so I could get an idea of the total fossil biota and an impression of the original ecosystem. I did not find a spider, but I did discover a rather nice leaf, and the journey through Otago to the little town of Middlemarch was lovely.

The scientific study of Foulden Maar is an excellent example how to do research on *Fossil-Lagerstätten*, which yield their secrets only after detailed scientific scrutiny. Each aspect of the geology and paleontology of this site has been investigated by experts, and Daphne Lee has tirelessly championed its international importance. The site's paleoecology provides data on food webs and ecosystems. Its entombing sediments offer insights into past climate patterns and can therefore be predictive for how ecosystems might alter as future climates change.

This beautifully written and illustrated book is both a comprehensive account of all that we know to date about this exceptional fossil locality, and a tribute to the scientists and journalists involved in revealing its secrets to the world. Like Grube Messel in Germany, which was rescued from becoming a rubbish dump in the 1970s and is now a UNESCO World Heritage Site, Foulden Maar needs to be conserved for research, education and tourism in perpetuity.

Paul Selden
Gulf-Hedberg Distinguished Professor Emeritus
University of Kansas

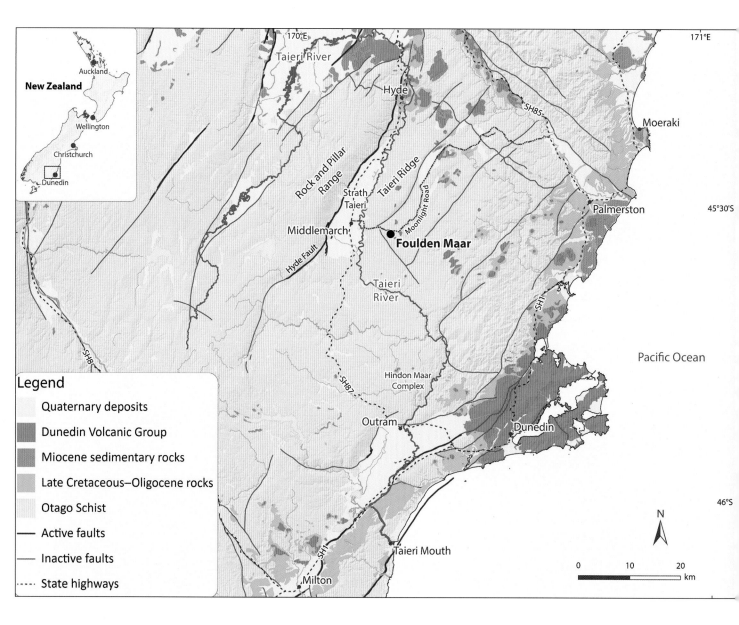

Legend

	Quaternary deposits
	Dunedin Volcanic Group
	Miocene sedimentary rocks
	Late Cretaceous–Oligocene rocks
	Otago Schist
——	Active faults
——	Inactive faults
----	State highways

New Zealand

Auckland

Wellington

Christchurch

Dunedin

Taieri River

Hyde

SH85

Moeraki

Rock and Pillar Range

Strath Taieri

Taieri Ridge

Moonlight Road

Palmerston

45°30'S

Middlemarch

Foulden Maar

Hyde Fault

Taieri River

Hindon Maar Complex

SH87

Pacific Ocean

Outram

Dunedin

SH1

SH8

46°S

SH1

Taieri Mouth

Milton

N

0   10   20
km

170°E

171°E

INTRODUCTION

Foulden Maar in Otago, New Zealand, is a paleontological site of international significance. Here a deposit of diatomite holds a unique record of terrestrial life in the mid-latitudes of the Southern Hemisphere. Formed by a volcanic eruption 23 million years ago (Ma), it comprises tens of thousands of undisturbed annual layers that record the changing life and ecosystems in and around a small, deep volcanic crater lake that existed for more than 130,000 years at the very beginning of the Miocene.

Plant fossils in Foulden Maar include millions of leaves, fragile flowers with pollen, fruits, seeds, wood and bark, together with pollen grains, spores and billions of diatoms. Animal fossils from the freshwater lake and surrounding rainforest abound and include the oldest known galaxiid fish on Earth, the first freshwater eel found in the Southern Hemisphere, sponges, and myriad insects and spiders. Fish larvae still have their patterned skin, insects their eyes, antennae and wing patterns, and some have retained their colour. Ecological interactions are also captured: scale insects may be seen in life position on leaves, fish have the remains of their last meal in their stomach, and there is insect damage on plants.

Most remarkable of all, Foulden Maar preserves a record of the climate fluctuations, year by year, through the life of the lake. Detailed analyses of core samples taken from the diatomite deposit have revealed temperatures, rainfall, global climate cycles and changing carbon dioxide (CO_2) levels in a world 23 million years ago. This unparalleled archive of past life and climate shows that the mean annual temperature in the area was 8°C warmer than today – a marginally subtropical climate. The increased temperature was connected to higher CO_2 levels of 450 parts per million (ppm) – approaching those that Earth will reach in the next few decades.

Foulden Maar today occupies a bowl-shaped area about a kilometre across in the midst of farmland not far from the township of Middlemarch. Livestock graze on flat paddocks that are typically greener than the surrounding rugged landscape. The other sign of something out of the ordinary is a series of small excavated pits filled with a white material.

To understand the story of Foulden Maar it is necessary to travel far back in geological time. Twenty-three million years ago a series of violent eruptions took place in the area. Hot basaltic magma rising through the Earth's crust hit a body of near-surface water at this site and exploded, creating a small, deep crater (a maar) and laying waste to the surrounding rainforest and landscape. Pulverised rock,

OPPOSITE Simplified geological map of East Otago. The basement rock shown in purple is the metamorphic rock known as Otago Schist. It is overlain in places by nonmarine and marine sedimentary rocks. Those of Late Cretaceous to Oligocene age are shown in green, and those of Miocene age are brown. Rocks of the Dunedin Volcanic Group shown in red are mostly Miocene in age, and they intrude into and overlie the older metamorphic and sedimentary rocks. Younger alluvial deposits of Quaternary age are coloured yellow. Foulden Maar is near the centre of the map, about 10km east of the township of Middlemarch. The northeast-trending asymmetric ridges of schist that make up Taieri Ridge and the much higher Rock and Pillar Range to the west are the result of uplift along reactivated reverse faults in the late Cenozoic. The Hyde Fault, which runs along the foot of the Rock and Pillars, is regarded as still active. SER

ABOVE Panorama from Bald Hill looking west across the near-circular crater of Foulden Maar with the largest open-cast pit of white diatomite near the centre. In the distance is the Strath Taieri backed by the Rock and Pillar Range under a blanket of cloud. UK

RIGHT The geologic time scale. Foulden Maar erupted close to the Oligocene–Miocene boundary, 23 Ma.

Adapted by Jenny Stein and Stephen Read[1]

An aerial view of Foulden Maar showing the main mining pit (Pit A), the location of the drilling site (FH2) and Holly Hill. The dashed line indicates the approximate edge of the maar. SER

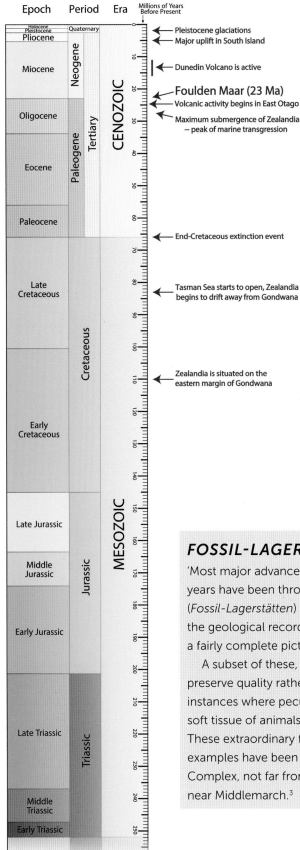

Epoch	Period	Era	Millions of Years Before Present	
			0	← Pleistocene glaciations

Epoch / Period / Era columns (geological time scale):

- Holocene / Pleistocene — Quaternary
- Pliocene
- Miocene — Neogene
- Oligocene — Paleogene / Tertiary — CENOZOIC
- Eocene
- Paleocene
- Late Cretaceous — Cretaceous — MESOZOIC
- Early Cretaceous
- Late Jurassic — Jurassic
- Middle Jurassic
- Early Jurassic
- Late Triassic — Triassic
- Middle Triassic
- Early Triassic

Annotations on the time scale:

- ← Pleistocene glaciations
- ← Major uplift in South Island
- | ← Dunedin Volcano is active (≈10)
- ← **Foulden Maar (23 Ma)**
- ← Volcanic activity begins in East Otago
- ← Maximum submergence of Zealandia — peak of marine transgression
- ← End-Cretaceous extinction event
- ← Tasman Sea starts to open, Zealandia begins to drift away from Gondwana
- ← Zealandia is situated on the eastern margin of Gondwana

sediment and ash rose in an eruption cloud, then fell back into the crater or was distributed by wind over the surrounding area. Volcanic debris deposited around the crater rim prevented streams from carrying sediment into the newly formed maar.

After the eruptions the crater rapidly filled with groundwater to form a deep isolated lake. Slowly, vegetation began to recolonise the devastated landscape and life invaded the lake. For thousands of years, microscopic single-celled diatoms and other algae bloomed on the lake surface each spring and summer. After death they slowly sank to cover the lakebed in a thin blanket of fine-grained siliceous ooze that over time built up to become diatomite. The bottom of the lake was anoxic and stagnant, so any plants or animals that drifted down were effectively pickled in the oxygen-deprived environment.

Eventually the crater was completely infilled. Much later, long after volcanic activity in the region had ceased, younger alluvial sediments covered the surface, capping and preserving this ecosystem time capsule for the next 20 million years or so.

In 1875 geologists searching for mineral resources reported a deposit of freshwater 'diatomaceous earth', as diatomite was then called, on the Strath Taieri east of the Rock and Pillar Range. From the 1940s to the

FOSSIL-LAGERSTÄTTEN

'Most major advances in understanding the history of life on Earth in recent years have been through the study of exceptionally well-preserved biotas (*Fossil-Lagerstätten*) ... Study of a selection of such sites scattered throughout the geological record — windows on the history of life on Earth — can provide a fairly complete picture of the evolution of ecosystems through time.'

A subset of these, *Konservat-Lagerstätten* (or conservation *Lagerstätten*) preserve quality rather than quantity. The term is restricted 'to those rare instances where peculiar preservational conditions have allowed even the soft tissue of animals and plants to be preserved, often in incredible detail'.[2] These extraordinary fossil sites are extremely rare globally, and only two such examples have been discovered in New Zealand. One is the Hindon Maar Complex, not far from the Taieri River gorge. The other is Foulden Maar, near Middlemarch.[3]

View from Moonlight Road across farmland, June 2009. The only indication that something unexpected lies beneath is the white pit on the right and the drilling rig in the middle of a paddock. DEL

1970s, a few truckloads of this diatomite were removed annually for commercial use, from four small surface mining pits.

Students and researchers from the University of Otago began visiting the site regularly from the 1960s to examine the unusual sediments and collect fossilised leaves. In 2003 a scientific research programme involving field mapping, volcanology, geophysical investigations, sedimentology and paleontology got under way, and scientists soon realised that they were uncovering fossil treasure with an extraordinary degree of preservation. In June 2009, a Marsden grant from the Royal Society of New Zealand enabled researchers to drill through the deposit. What they found confirmed that this was indeed a maar crater.

Scientific research continued in parallel with small-scale diatomite mining for some years. But in

ABOVE A general view of diatomite in the main pit. A few metres of geologically young alluvial gravels sit on top of a layer of weathered white diatomite. **LEFT** Removal of a thin yellowish surface veneer reveals the dark, finely laminated diatomite.
DEL

Students hunt for fossils in the main pit, 2011. UK

April 2019, plans for a massive expansion of mining operations were revealed. For the first time it dawned on scientists and the local community that this unique and globally important site could be destroyed. With unparalleled scientific findings now emanating from Foulden Maar, decisive action was necessary to protect this – and probably other significant fossil sites – from imminent destruction.

The resulting 'Save Foulden Maar' campaign included a petition that garnered more than 10,000 signatures in two months. In an effort to alert the public to this situation, campaigners composed hundreds of emails and letters, held media interviews, wrote articles, gave dozens of public talks, made submissions to the Overseas Investment Office and staged art and poetry exhibitions.

In June 2019, just seven weeks later, the company

Slabs of diatomite (**LEFT**), blocks of diatomite with leaves (**MIDDLE**) and curled-up diatomite layers after exposure to the atmosphere (**BELOW**). The dense, dark diatomite beneath the water table has a high water content. As it dries out the layers separate and become more brittle, resembling dried cardboard in appearance and texture. The colour gradually changes to pale grey and, after long weathering, becomes pure white. Geological hammer for scale. DEL, JKL

that proposed to mine and export all the diatomite at Foulden over a period of 27 years was placed in voluntary receivership and went into liquidation. For the time being, the immediate threat was removed.

Worryingly, however, this globally important unique fossil site is still at risk of destruction because no legislation exists in New Zealand to protect geological and paleontological taonga/treasures on privately owned land. If it were an archaeological site, a wetland or a site with living biodiversity values, then protection might be possible. But the ecological biodiversity of New Zealand's past currently has no legal status.

As this book goes to press, options are being negotiated to protect Foulden Maar and all its treasures for future scientific and educational purposes.

This book explores Foulden Maar and celebrates its treasures.

RIGHT Simplified schematic geological/ stratigraphic column to show the sequence of rock types present at Foulden Maar. UK

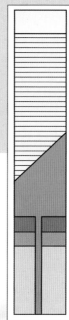

Quaternary Alluvium (<2.6 Ma)

Foulden Diatomite
Early Miocene (23 Ma)

Dunedin Volcanic Group
Late Oligocene to late Miocene (25–9 Ma)

Pakaha Group
Late Cretaceous to Paleogene

Otago Schist
Early Jurassic to early Cretaceous metamorphism (~200–115 Ma)

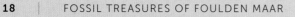

CHAPTER 1
THE GEOLOGICAL CONTEXT

Foulden Maar has a complex geological setting and a long history. It is situated in a region of East Otago where long-extinct volcanoes have erupted through the basement metamorphic schist, where the sea came in briefly and then receded, and around which the mountains are still rising along active faults.

OTAGO SCHIST

The landscape around Middlemarch is dominated by Otago Schist with its distinctive free-standing craggy rocks or tors. This platy, mica-bearing, often quartz-veined and occasionally gold-bearing rock forms the geological bedrock or basement throughout much of Otago. Schist is a metamorphic rock formed by the deep burial, heating and deformation of older rocks, in this case mostly Triassic marine sandstones and mudstones and minor amounts of volcanic rocks.

During the Jurassic period they were buried to depths of up to 15km, heated to about 400°C, compressed and folded, and in early to mid-Cretaceous times these metamorphic rocks were uplifted to the Earth's surface as Otago Schist. Over millions of years in the late Cretaceous, about the time Zealandia was separating from the rest of Gondwana, they were eroded to a low-elevation but widespread and almost level 'peneplain'. This erosion surface across the top of the schist is present over most of Otago.[1]

YOUNGER ROCK LAYERS: THE PAKAHA GROUP

Tens of millions of years after the formation of the erosion surface on the schist, younger sedimentary rocks began to be deposited on top. In the Middlemarch area these sediments are collectively called the Pakaha Group (coloured green on the geological map).[2]

Most is sand and gravel deposited by ancient rivers, but at Slip Hill, a few kilometres northwest of Foulden Maar, a thin layer of quartz sands containing the marine mineral glaucony and marine fossils (foraminifera and shell fragments) rests on or near the schist basement.[3] This represents the remnants of a marine transgression that extended over large parts of eastern Otago in the late Eocene and Oligocene. These shallow marine sands are not exposed at Foulden, but grains

OPPOSITE In fresh outcrops, schist is grey in colour, well foliated (it splits along flat planes) and contains abundant 1mm–5cm-thick quartz veins parallel to foliation (**INSET**). Around Foulden the schist foliation dips at low angles to the northeast and east. UK

Silcrete boulders (sometimes called sarsen stones) are residual silcrete layers formed within fluvial-channel quartz sandstone and pebble conglomerates. They are common throughout Otago. JKL

of quartz sand and glaucony found in the sediments associated with the maar suggest they may have been present at the eruption site.

Silica-cemented schistose quartz–pebble conglomerate and sandstone boulders are scattered across paddocks near the maar. Some are isolated silcrete boulders and others form 10–30m-wide boulder fields. In the absence of fossil evidence their provenance is uncertain: they may be of Eocene age, resting on the schist basement, or Miocene age, postdating the volcanics.

VOLCANIC ACTIVITY

Eastern Otago was once a hotbed of volcanic activity. The eroded remnants of volcanoes are scattered throughout the area and contribute to the complex and distinctive landscape features around Foulden Maar.

Though schist forms the basement rock and thin layers of sediment are sometimes preserved above it, the icing on the sedimentary layer cake is made up of volcanic rocks of many types: scoria cones, dikes, lava plugs, hill-capping basaltic flows. Shown in red on the geological maps, these are mostly dark-coloured and fine-grained, sometimes with large gas bubbles or vesicles.[4]

Foulden Maar is one of about 150 volcanoes within the Dunedin Volcanic Group (DVG), which includes the Waipīata Volcanic Field. Previously regarded as separate, the Dunedin and Waipīata fields have been shown to overlap in time and space.[5] The DVG covers an extensive area of about 7800km² from Gimmerburn, near Ranfurly, across to Palmerston in the north, south to Milton and east to the Otago coast. It is one of the largest intraplate volcanic provinces in New Zealand.

Eruptions began near Middlemarch about 25 Ma and became regionally widespread. The city of Dunedin is built on the younger Dunedin Volcano, by far the biggest edifice within the DVG and active from 16 to about 11 Ma. Volcanic activity had ceased by 9 Ma and the DVG is now regarded as extinct.

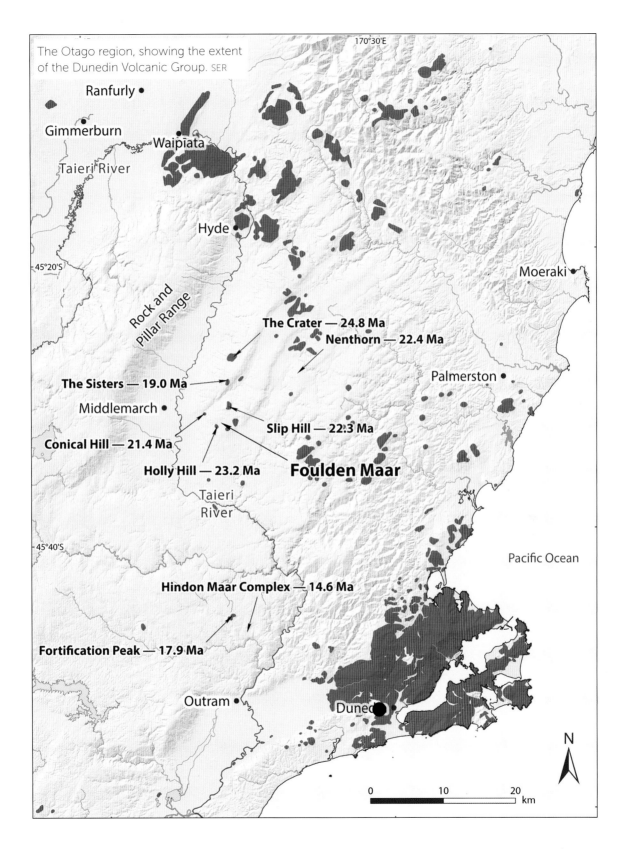

The Otago region, showing the extent of the Dunedin Volcanic Group. SER

170°30'E

Ranfurly •

Gimmerburn •

Waipiata •

Taieri River

Hyde •

45°20'S

Moeraki •

Rock and Pillar Range

The Crater — 24.8 Ma
Nenthorn — 22.4 Ma

The Sisters — 19.0 Ma

Middlemarch •

Palmerston •

Slip Hill — 22.3 Ma

Conical Hill — 21.4 Ma

Holly Hill — 23.2 Ma   **Foulden Maar**

Taieri River

45°40'S

Pacific Ocean

Hindon Maar Complex — 14.6 Ma

Fortification Peak — 17.9 Ma

Outram •

Dunedin

N

0   10   20
km

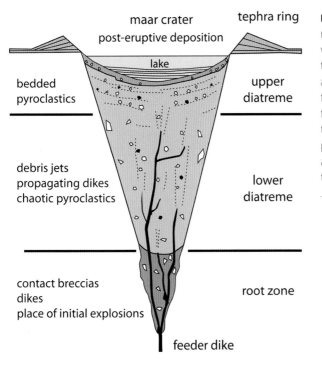

maar crater — tephra ring

post-eruptive deposition

lake

bedded pyroclastics

upper diatreme

debris jets
propagating dikes
chaotic pyroclastics

lower diatreme

contact breccias
dikes
place of initial explosions

root zone

feeder dike

LEFT Cross-section of a maar-diatreme volcano with a maar lake, showing the feeder dike below an irregular root zone, a funnel-shaped diatreme filled by pyroclastics and the maar crater filled by post-eruptive sediments and surrounded by a tephra ring. UK

DATING OTAGO'S VOLCANISM

Exactly when the volcanic activity occurred in the area around and including Foulden Maar has been determined by radiometric dating methods that calculate the age of rocks using radioactive isotopes. These show that volcanic activity began about 24.8 Ma and continued intermittently until around 9 Ma. In contrast to older volcanoes around Ōamaru and Kakanui in North Otago, which were generally submarine, most erupted on the surface. The earliest eruptions were at The Crater, a conspicuous landmark on Taieri Ridge. The next were closer to Foulden Maar: Holly Hill on the edge of the maar is dated at 23.2 Ma. Nearby volcanic features such as Conical Hill and Slip Hill have been dated at 21.4 and 22.3 Ma respectively.[6]

LEFT Two kilometres southeast of Foulden Maar, Conical Hill (locally called Smooth Cone) resembles the conventional image of a volcano that has erupted through the schist basement. However, its cone shape is actually the eroded remnant of the conduit of a former volcano (**ABOVE**). DEL, UK

LEFT AND ABOVE At Holly Hill, on the edge of the Foulden Maar crater, a 25-metre-long, 2–4 metre-wide dike of fine-grained volcanic rock is exposed on the surface. Dark grey to black in colour, this basanite has conspicuous platy jointing and large circular vesicles that formed as gas in the lava bubbled off. Because the rock is unweathered and contains suitable mineral grains, a sample from Holly Hill was selected for radiometric dating. JKL

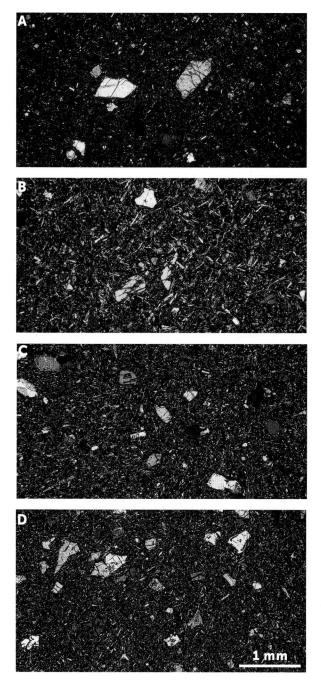

Photomicrographs of fine-grained volcanic rocks from the vicinity of Foulden Maar, seen under cross-polarised light. **A** from Bald Hill; **B** from Holly Hill; **C** from Conical Hill; **D** from Slip Hill. The fine-grained groundmass is mostly feldspar; the larger crystals are olivine and augite.[7] UK

WHAT IS A MAAR?

A maar is a type of volcano with a crater cut into the pre-eruptive landscape, surrounded by a tephra ring, and underlain by a diatreme. It forms when magma, heated to 800–1200°C, rises to the Earth's surface and comes in contact with wet sediment or water in an aquifer, stream or lake. The hot/cold mix creates violent steam explosions and a huge plume of steam and pulverised rock that rises high in the air.

Most of the debris from the eruption column falls back to partially fill the crater, or forms a rim of tephra – loose pyroclastic material from the explosive volcanic eruption – around the crater's edge. A blanket of ash may extend for tens of kilometres from the crater, depending on the prevailing wind. Once the eruptions cease, a deep steep-sided pit remains – a maar crater.

Maars are among the most common volcanic features in continental volcanic fields, but unlike volcanic cones they are often hidden landscape features and may not be recognised as volcanoes. Many people in New Zealand have never heard of them, yet this country is host to dozens of these craters, many of them in the Auckland Volcanic Field (Lake Pupuke is one).[8] Foulden Maar is a maar-diatreme volcano, a reference to the long, vertical pipe through which gas-filled magma rises to the Earth's surface during the eruption.

Maars vary in diameter from 100m to 3000m, and most occur in magmas of basaltic composition. There is a relationship between diameter and depth, and both depend on rock and magma types, the hydrology, and the number and duration of the eruptions.[9] Maars also sometimes overlap and coalesce, and may form a series, like beads on a string.

Maar eruptions commonly last only a few hours or days as they are fuelled by contact between magma and water, and few have been witnessed. In March–April 1977 the Ukinrek Maars erupted in a remote part of Alaska creating two small craters, one 170m in diameter and 35m deep, the other 300m across and 70m deep. Other geologically recent maar eruptions have taken place at Nilahue in Chile (1956) and Lake

Taal in the Philippines (1965). New Zealand's 1886 Tarawera eruption was also maar-forming.[10]

The term 'maar' comes from the German *Meer*, which in turn derives from the Latin *mare* (sea). Maar craters commonly undercut the local groundwater level and fill with water to become a lake. Circular to oval in shape, a maar lake is a closed system, its catchment restricted to the inner slopes of the surrounding tephra rim. As a result, little land-based sediment and only small amounts of wind-blown dust get incorporated in the lake system. Instead, the type of sediment and the annual sedimentation rate are mainly a function of the bioproductivity within the lake, which in turn is controlled by the lake's hydrology, the local environment and the climate.

The deepest lakes are usually meromictic, meaning they have a well-mixed oxygenated upper water layer

The Pulvermaar in West Eifel Volcanic Field, Germany, is a typical geologically young maar of Pleistocene age. This lake is ~700m in diameter (smaller than Foulden) and 72m deep. Note the low, forested tephra rim around the crater. Courtesy of Maar Museum, Manderscheid/ H. Gassen

and a lower stagnant water column that is depleted of oxygen (anoxic), rich in dissolved minerals and free of benthic biota – bottom-dwelling plant or animal life – that might disturb the lakebed. Biogenic sediment produced from algae and other organisms living in the water drifts to the lakebed and settles. Season by season, these thin deposits of dead algae typically alternate with layers of organic detritus, such as leaves from the surrounding vegetation and minor amounts of clay or silt washed in from soils and vegetation on the tephra ring. In volcanically active regions, a maar lake may also capture layers of ash from nearby eruptions, which can be useful for dating the lake sediments or establishing a chronology of volcanic events in the region. Over time – perhaps 200,000 years – a deep lake may accumulate up to 200m of sediment.

Maar lake sediments have a high potential for preservation because they lie below the landscape surface. Examination of these sediments can provide unique and invaluable information about the paleoenvironment, paleoclimate, landscape evolution and erosion history of a region. Only a few near-complete maar crater sequences have been studied globally. Foulden Maar provides one of these sequences.

GEOPHYSICAL INVESTIGATIONS AND SCIENTIFIC DRILLING

The sediment that has infilled the Foulden Maar crater is formally described as Foulden Hill Diatomite. Geological mapping shows that the deposit covers a roughly circular area with a diameter of about a kilometre. Exposed diatomite in four mining pits, some dating back to the 1940s and 50s, and some short drillholes made in the 1950s initially suggested the deposit was 20–40m thick.

The mining pits are shallow open-cast excavations that extend about 10m below the ground surface. Thousands of light- and dark-coloured layers are readily visible in the pit walls. These are biogenic varves – thin layers that represent a single year's sediment. Each individual layer is a bedding plane, laid down parallel

One of the most striking features of the diatomite is the conspicuous layering. Jon Lindqvist used a chainsaw to cut out vertical columns of diatomite, then carefully photographed, measured and counted the dark- and light-coloured layers exposed in the two main pits. DEL, JKL

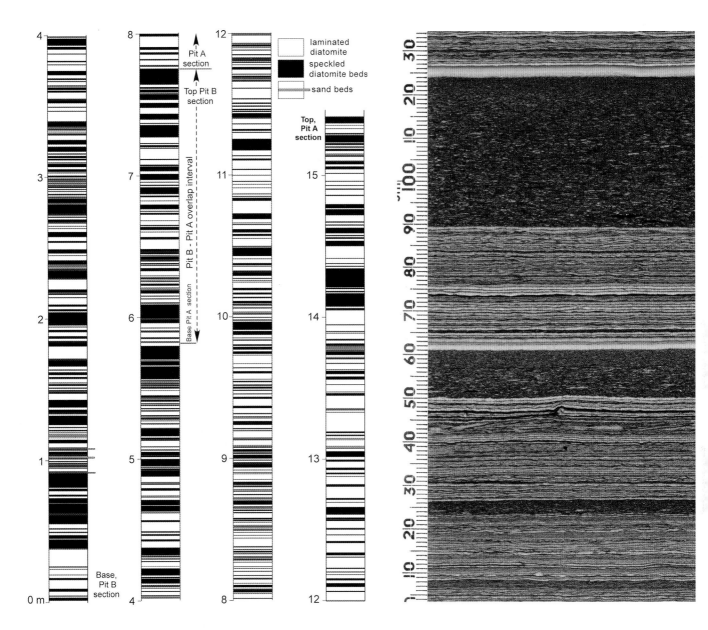

to the original surface of deposition that was once the actual lake floor. Annual varves occur in couplets, each differing slightly in thickness and colour. Those at Foulden range in thickness from 0.25 to 2mm. The varves can be traced and matched up across the deposit for tens to hundreds of metres, effectively 'barcoding' the sediment layers.

These layers were one of the first clues that this was indeed a maar, since such biogenic varves are found only in deep closed-system lakes. A second clue was the

ABOVE LEFT The layers were used to correlate the sections exposed in the pits with the upper section of the core. **ABOVE RIGHT** In addition to paired varves, there are 'speckled beds' or dark diatomite beds up to 150mm thick. These incorporate pale diatomite flecks, fine carbonaceous material and leaves, and invariably have a 1–5mm-thick capping layer of white diatomite.[11] These beds were deposited at intervals of hundreds of years, from sediment gravity flows generated from the down-slope collapse of plant matter-rich diatom ooze accumulating around the lake margins. JKL

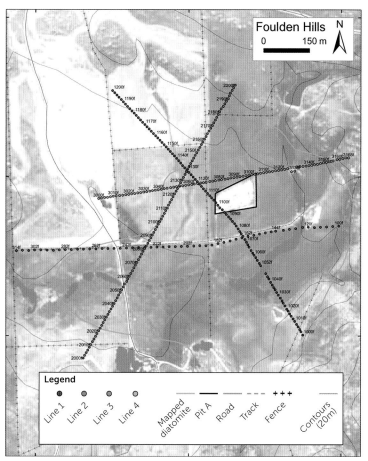

Foulden Hills
0   150 m
N

Legend

● Line 1   ● Line 2   ● Line 3   ● Line 4   —— Mapped diatomite   —— Pit A   —— Road   —— Track   +++ Fence   —— Contours (20m)

ABOVE University of Otago Geophysics Field School students collect gravity and seismic data across the maar surface in 2006. Andrew Gorman

LEFT Seismic reflection imaging is the best method for determining the structure of the sedimentary deposits in a maar crater such as Foulden. To achieve this, dozens of shots of explosives were set off in a carefully arranged pattern in 2006 and again in 2008. Daniel Jones

near-circular basin in the farmland that shows green in dry seasons – from the water-saturated sediment beneath – when the surrounding landscape is brown.

Final proof that this was a maar could only come from drilling right through the thickest part of the deposit to reach the volcanic diatreme beneath. The drilling process is expensive, so it was essential to establish the sub-surface structure beforehand. Staff and teams of geophysics students from the University of Otago Geology Department got to work carrying out seismic, gravity and magnetic surveys. These used the physical properties of rock types, such as density and magnetism, to determine what lies hidden beneath the surface.

The seismic surveys used dozens of small-scale controlled explosions in transects across the surface of the bowl-shaped area to generate waves that resemble those produced in small earthquakes. The waves are reflected and refracted off the layers beneath the surface and help to build up a picture of the shape of the underlying structure.

Maars also commonly display a circular low-gravity anomaly. The Earth's gravity field is affected by the density of different rock types: sedimentary rocks are generally less dense than volcanic rocks or metamorphic rocks such as schist. Diatomite has a particularly low density, which was measured directly from the pits. Gravity measurements made from 356 stations over eight transects showed that there was indeed a low-gravity anomaly, suggesting a thickness of about 120m of diatomite near the centre of the structure.

Researchers measured magnetic anomalies by walking back and forth across the surface of the presumed maar with a magnetometer, recording the

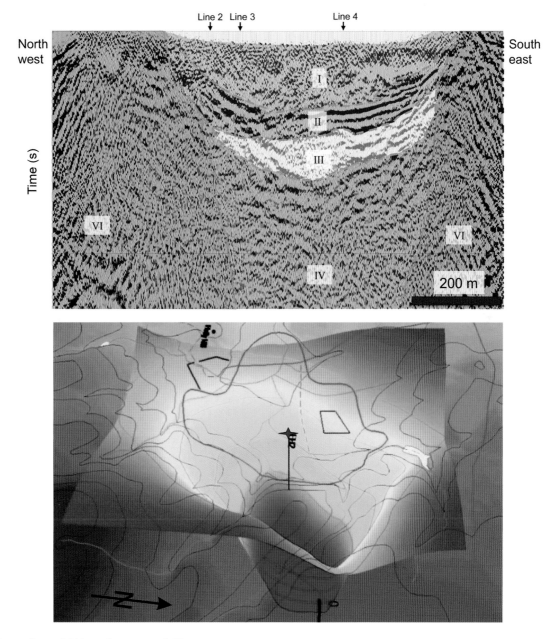

magnetic intensity at 2011 stations spaced 20m apart across a grid pattern. Volcanic rocks such as basalt contain the mineral magnetite, which, as the name suggests, is highly magnetic. Results showed a high-intensity magnetic field beneath the centre that was likely to be caused by a volcanic root structure.[12] After much processing and analysis, a 3D model of what lay beneath the surface was constructed and a drilling target chosen.

The Foulden Maar subsurface structure was worked out prior to drilling by geophysical investigations of gravity, magnetism and seismic reflection. Daniel Jones produced the first 3D virtual image of the maar as part of his MSc research. The dark blue indicates the thickest amount of sediment. Daniel Jones

In June 2009 a team from Webster Drilling arrived on site with a rig that was to be trialled in midwinter in Otago before going south to Antarctica. With the ground frozen on some days and snow falling on others, the weather was suitably Antarctic-like.

Drilling through soft sediment such as diatomite is straightforward, but retrieving a complete core is not. The team drilled two 85mm-diameter cores: FH1 to a depth of 120m, and FH2, 10 metres away, to 183m. Any core loss at one site could then be spliced in from the other, thanks to the 'barcoding' of the sediment.

Altogether about 300m of core was successfully retrieved, labelled and taken back to be stored in a refrigerated container at the University of Otago, where it would be analysed using a Geotek Multi-Sensor Core Logger in the Geology Department laboratory.

The 183m-long section is one of the few complete cores to have been retrieved from the sediments of a pre-Quaternary maar lake anywhere in the world.[13] From its base to its top (the oldest layers to the youngest), four distinct lithological zones can be

TOP Daniel Jones and Beth Fox handle sections of core. DEL

ABOVE A selection of labelled core boxes. DEL

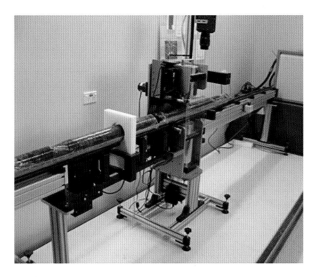

The Geotek Multi-Sensor Core Logger. A section of the core is wrapped in clingfilm and laid on a slowly moving bed. Rock properties such as P-wave velocity, gamma density and magnetic susceptibility are recorded. Later, the core is split in half and scanned with a high-resolution camera to provide images of the different rock types, the nature of their contacts and RGB (red, green, blue) colour values of the laminated diatomite for statistical analysis. UK

OPPOSITE The core tells the story of Foulden Maar from the formation of the crater to the present. These examples of short sections from the base to the top of core FH2 show the typical appearance of each of the four lithozones. UK

distinguished, based on the prevailing rock types and their geophysical properties.

Lithozone A (opposite) is the section from 183.8m to 177m, the deepest part of the core. It consists of massive unsorted breccia – angular fragments of schist basement and basaltic lapilli (volcanic rock fragments) in a matrix of mud and volcanic ash. This is interpreted as deposits resulting from early post-eruptive collapse of the crater walls, along with material from the tephra rim surrounding the newly formed crater. Although only around seven metres of this is apparent in the bottom of the core, the total thickness of the breccia section is probably a few tens of metres: drilling was stopped at 183.8m, soon after reaching this very hard rock.

Lithozone B is the 62m-thick section from 177m to 115m that comprises mudstone and sandstone beds,

unsorted carbonate-cemented schist breccia, pre-eruptive siliciclastic sediment, pyroclastic material (rocks ejected during an eruption) and fragments of basalt. These widely ranging fragmentary rock types are derived from the pre-existing rocks at the eruption site and debris from the tephra rim, and would also have originated as rockfalls or debris flows from the steep crater walls and the unstable rim following groundwater inflow and flooding in the central part of the crater.

Lithozone C, 115m to 102m, is characterised by alternating layers of diatomite, schistose and/or volcanogenic sand and breccias containing pebble- to cobble-sized schist or basalt clasts and rare redeposited volcanic ash. This is interpreted as background lake sedimentation interrupted often by underwater mass-flow events from the tephra rim. Rare rounded quartz and occasional marine-origin glaucony grains in some coarser layers indicate that a marine sandstone cover was present in the Foulden area at the time of the eruption.

Lithozone D extends from 102m to 6m and includes the 120m-thick uppermost section and consists mainly of two depositional types. One is a thinly laminated diatomite in pale and dark couplets averaging 0.5mm in thickness; the second consists of beds of laminated diatomite clasts in a fine-grained diatomaceous matrix. The uniform composition indicates long-lasting stable and calm conditions in the lake.

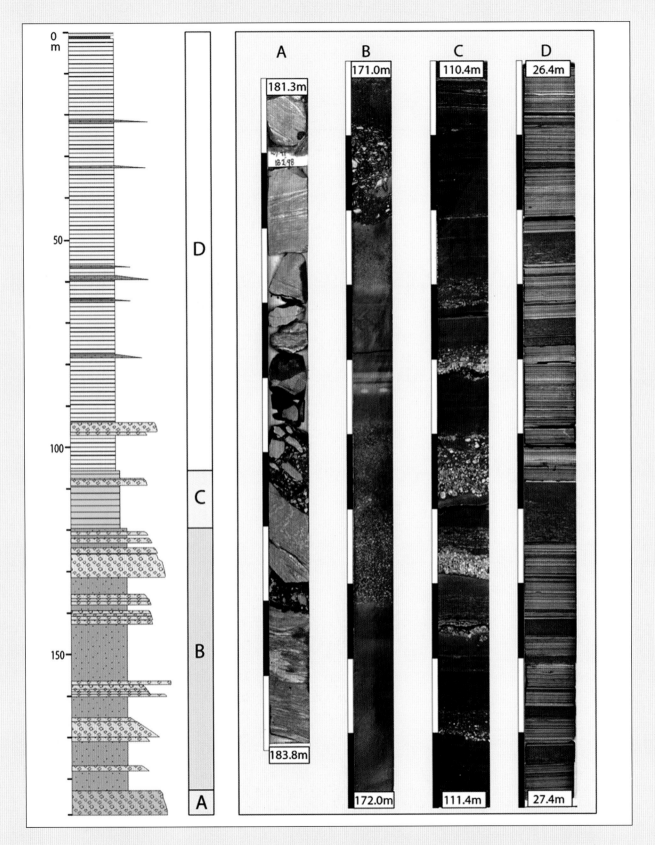

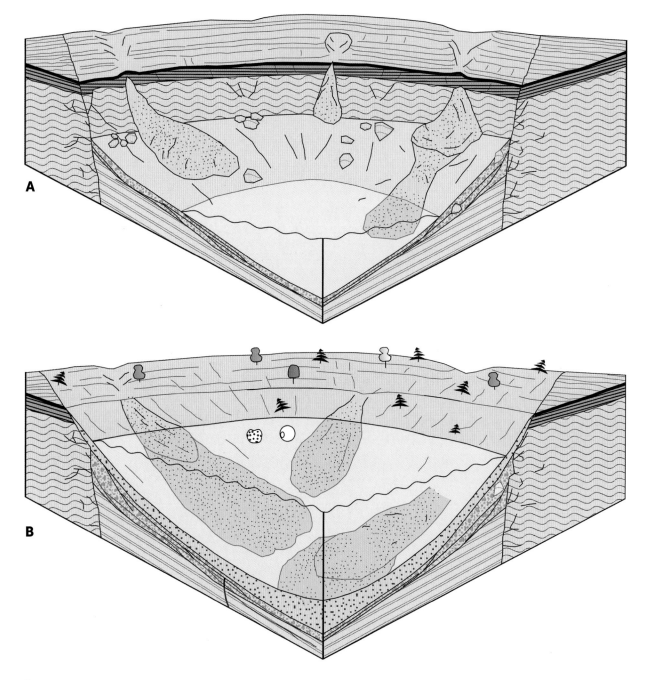

The post-eruptive history of Foulden Maar reconstructed from drill core data, geophysics and fossils. UK

A Days to months after the maar-forming eruptions, collapses of the steep crater walls and inner tephra rim deposit blocks of schist and volcanic debris on the floor of the fresh crater. Groundwater discharge has formed an initial maar lake.

B Decades after the eruption, the lake level has risen. Debris flows continue to carry loose material from the tephra rim into the deeper parts of the crater. Green algae, golden-brown algae and freshwater sponges are the first colonisers of the maar lake. A sparse pioneer vegetation of mainly ferns has started to grow on the steep rim.

C After decades to centuries, forest plants have colonised and stabilised the partly eroded tephra rim. Diatoms

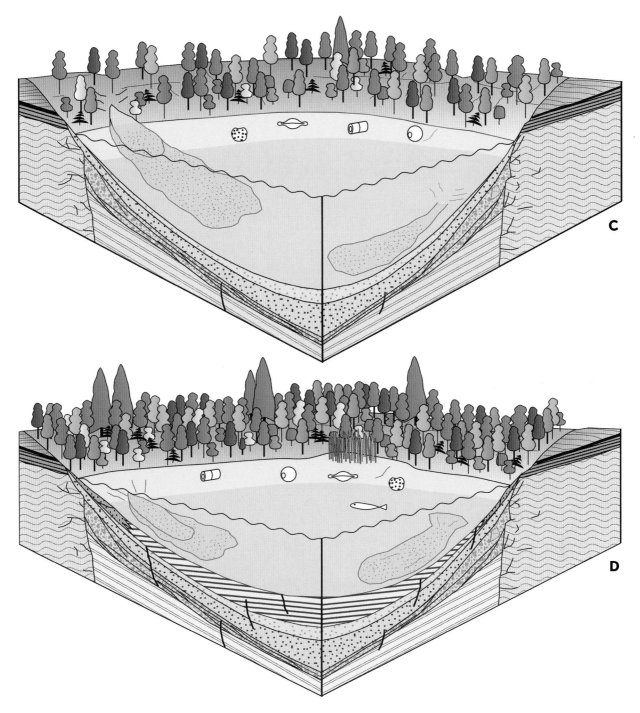

flourish in the upper water column of the now stratified maar lake, forming laminated layers on the lake floor after seasonal blooms. The bulk of sediment deposited, however, is sand and gravel carried in from the surroundings by mass flows.

D For over 100,000 years, seasonal algae blooms alternating with wash-in of organic debris in wet seasons have accumulated as laminated sediment on the anoxic lake floor, occasionally interrupted by coarse-grained mass-flow deposits. Volcanic soils have formed on the tephra rim and support a dense rainforest that clothes the lake margin. Insects are abundant in the forest, and some aquatic insect species live in the upper water column, together with freshwater sponges, algae, fish and water plants.

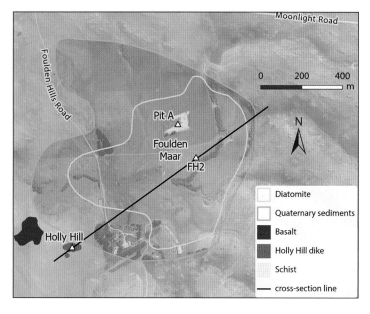

Summary map of geological and geophysical features present at Foulden Maar and a vertical cross-section to show what lies beneath the surface. Outcrops of volcanic rocks, including the dike at Holly Hill, are shown in red and black. SER (modified from Jones et al. 2017)

HOW OLD IS FOULDEN MAAR?

The date of the eruption that created Foulden Maar is put at 23.03 Ma, a critical era in Earth's history at the boundary between the Oligocene and Miocene epochs and a time of considerable relevance for predicting future climate changes (see Chapter 7).[14]

Two methods were used to discover when Foulden Maar erupted and thus the age of the lake sediments that accumulated in the maar. The first was based on fossil pollen studies. New Zealand has a long-established biostratigraphic framework that is used to date sedimentary rocks, based on the first and last appearances in the fossil record of key taxa, including pollen types. Geologists in the 1960s considered the diatomite deposit to be of Pliocene age (5.3–1.8 Ma). However, more recent pollen research based on numerous terrestrial fossil sites in Otago and Southland has confirmed that the diatomite is earliest Miocene in age (23 Ma).[15]

RIGHT The cross-section passes through Holly Hill and FH2 where the core was drilled. Features to note are the root zone, the unstratified diatreme breccia with blocks of the schist basement rock, the feeder vent or dike for the rising magma, the collapse breccia, debris flows and laminated diatomite that infilled the lake. Note also the present-day topography. The original tephra ring that would have surrounded the maar has been completely removed by erosion.[16] UK (modified from Jones et al. 2017)

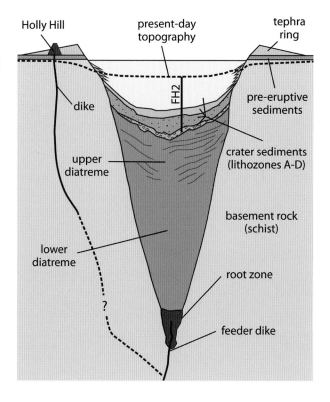

The second method used radiometric dating. This uses the natural decay of radioactive isotopes of argon in the minerals present in a sample, and is regarded as one of the most precise methods of establishing the age of a volcanic rock. In 2009 samples of basanite, the volcanic rock from Holly Hill, were sent to a laboratory in Kiel, Germany, which provided a radiometric date of 23.17 ± 0.19 Ma.[17] Once the 183m core had been retrieved from Foulden Maar, volcanic clasts from 126m depth were sent to a laboratory in the US to be dated, and one of these yielded a date of 23.38 ± 0.24 Ma – very close to the Holly Hill date. Another clast returned an older date of 24.51 ± 0.24 Ma, suggesting it had originated from an earlier eruption in the vicinity.

One other key (and unexpected) piece of evidence to support this date came from the discovery of a magnetic reversal in the core at a depth of about 107m. Earth's magnetic field reverses its polarity about once every 200,000 years, although this is random and highly variable: the last time it happened was in fact 780,000 years ago. Volcanic rocks and sediments containing magnetic minerals record the reversals, which have been captured in a Global Magnetic Polarity Time Scale that covers the past 60 million years. As there were few reversals in the relevant time period, the discovery of the reversal at Foulden Maar provided a further, independent method of dating the lake sediments.[18]

FOULDEN: A MODEL MAAR

Foulden Maar conforms to the typical funnel-shaped structure of maar volcanoes. The components in the basal crater sediments indicate that Foulden Maar erupted into a mixed hard-soft rock substrate of metamorphic Otago Schist overlaid by a thin deposit of weakly consolidated (now eroded) glauconitic sands and fluvial sediments.

The longer a maar is active, the greater the diameter of the crater and the depth of its diatreme. Prior to erosion, the crater at Foulden may have been as large as 2500m across and 350m deep. Unfortunately no trace of the surrounding tephra rim remains to verify this. Originally horizontal, the beds within the maar can be seen to dip towards the centre of the crater. This is typical of most maar deposits and reflects compaction of the breccia at the base of the sediments. Correlation between the sedimentary layers in the core and those visible in the walls of the mining pits adds a further 10 or more metres to the record, making the diatomite at Foulden at least 130m thick. And since it has been estimated that some 120m of erosion has occurred in the vicinity, the diatomite may have originally been about 250m thick.[19]

The varve layers in the centre of the crater show no break in sedimentation – no evidence of the lake drying out or an influx of land-based sediment from an in-flowing stream – so the lake may have remained a closed system throughout its lifespan.

Although Foulden Maar was eventually capped with younger alluvial sediments, it has not been subjected to deep burial by overlying sedimentary or volcanic rocks, which accounts for the pristine nature of its fossils and its remarkable climate record. The deep basin hidden beneath the surface was below the groundwater table and thus remained water-saturated until mining began in the 1940s, and this also helped preserve the diatomite and fossils hidden within it.

At the end of its life the lake would have become increasingly shallow, then a marshy swamp and eventually dry land. In the Eifel region of Germany, where maars were first studied, it is possible to see such a sequence. Visitors can walk from a water-filled lake to a swamp, to spongy ground and finally have lunch in a café built on solid ground above a completely dried-up former maar.

Frederick Hutton (1836–1905), the second (and last) Otago provincial geologist, was based in Dunedin 1873–80.

E. Wheeler and Son, 'Making New Zealand: Negatives and prints from the Making New Zealand Centennial collection', MNZ-0474-1/4-F, Alexander Turnbull Library, Wellington

Mineralogist George Ulrich (1830–1900) became first director of the School of Mines at the University of Otago.

Photo F. Hasler, Box-032 Port1602, Hocken Collections Uare Taoka o Hākena, University of Otago

Robert Speight (1867–1949), assistant curator (1911–14) and curator (1914–35) of Canterbury Museum.

Photo by Steffano Francis Webb, Original photographic prints and postcards from file print collection, Box 13, PAColl-6407-63, Alexander Turnbull Library, Wellington

CHAPTER 2
DIATOMITE AND MINING AT FOULDEN MAAR

Mr R. Speight, assistant curator at Canterbury Museum, has received from Middlemarch, Otago some interesting specimens … They are pieces of diatomaceous earth, which it is reported came from a deposit covering about 120 acres, with a thickness of 70ft. Mr Speight states that if this is so the deposit will evidently rank as one of the most important in the world.

— OTAGO DAILY TIMES, *6 July 1910*

Mining of many commodities, particularly coal and gold, has featured large in the history of the Province of Otago since it was founded in 1853. Following the discovery of gold in 1861, the Otago Provincial Council established the Otago Geological Survey. James Hector was appointed as the first provincial geologist and given the task of exploring and mapping the province's geological resources. Hector moved to Wellington in 1865 to set up the New Zealand Geological Survey, and in 1873 Frederick Hutton, a paleontologist from England, became his replacement in Otago.[1]

Hutton and mineralogist George Ulrich made the first mention of 'a deposit of diatomaceous earth' on the Strath Taieri in their *Report on the Geology and Gold Fields of Otago* in 1875.[2] It is not clear whether either man had visited the site but they had certainly seen a sample of the material. This may have come from a creek bed, as there is no natural surface exposure.

Microscopic examination revealed diatoms. The geologists (presumably Hutton) wrote that the sample 'is chiefly made up of three or four minute species of *Cymbella* or *Cocconema*', and 'although I have not been able to determine any of them, I have no doubt but that they are freshwater species'. The report later mentions the occurrence of 'polishing power [powder] diatomaceous earth in Strath Taieri'.[3]

The large Cottesbrook Run, where the diatomite had been found, was subdivided into smaller farms in 1885. The Thompson family bought the portion in question and named it Foulden Hill (often mistakenly rendered Foulden Hills) after a site with family connections near the England–Scotland border.[4] Foulden Hill remained in the Thompson family for over a century until it was sold in 1992. It has had several owners since then, all of whom have continued the pastoral farming tradition.

After this brief mention, it was another 35 years before the 'diatomaceous earth' on the Strath Taieri again appeared in print, this time in relation to its economic potential. The *Otago Daily Times* report at the opening of this chapter gave the impression that this was a new discovery. It quoted Robert Speight of Canterbury Museum, who stated that 385 bags of the material had been sent to England for analysis. His comment that the deposit would 'rank as one of the most important in the world' was remarkably prescient, though not for the reasons he had in mind.

DIATOMITE AND ITS USES

Diatomite is a sedimentary rock consisting principally of minute siliceous aquatic algae called diatoms. Most diatomite is marine in origin. Today it is heavily mined for use mostly in filtration processes, where it purifies fluids by aggregating and collecting suspended solids. It is also used as an absorbent, such as in pet litter, and as a lightweight aggregate in concrete. Vast open deposits are currently processed in the US, Denmark, Turkey, China, Peru, Mexico and elsewhere. According to the United States Geological Survey, there are adequate diatomite resources to meet global demand for the foreseeable future.[5]

Diatomite deposits of both marine and freshwater origin are common throughout New Zealand, often associated with volcanic rocks. The 1927 report, 'Minerals and Mineral Substances of New Zealand', listed dozens of sites from Northland to Otago where diatomite or diatomaceous earth had been found. According to the report, it had many industrial uses including 'as an absorbent for nitro-glycerine in the manufacture of dynamite; as a heat-insulator for covering boilers and steam-pipes'.[6] It was used in fireproofing, polishing powders, pottery, bricks, filtration and surgical dressings.

A more detailed paper by geologist L.I. Grange in 1930 mentioned further diatomite deposits near Whangārei, at Kingsland and Takanini in Auckland, at several sites near Rotorua and Taupō, at Craigieburn in Canterbury and Frasers Creek and Middlemarch in Otago. This paper also included a chemical analysis for the Middlemarch diatomite: 76.42% silica, 5.32%

alumina, 1.92% ferric oxide, 0.36% magnesia, 1.00% lime and 14.10% water and organic matter.[7]

The first scientific study of the Middlemarch diatomite was carried out by Ray Gordon in the 1950s. His unpublished report was commissioned by the Ministry of Works in 1954, and was followed in 1959 by two presentations at the Fourth Triennial Mineral Conference held at the School of Mines and Metallurgy at the University of Otago, Dunedin.[8] At the time, the large Roxburgh and Waitaki hydroelectric dams were under construction in the region, and the diatomite was being considered for use as pozzolan, a hydration and strengthening agent in the concrete. Gordon established the extent and approximate depth of the diatomite deposit using boreholes, calculated the reserves, and projected the likely cost of mining and transportation from Middlemarch to Roxburgh. In the end, however, only a few hundred tonnes of diatomite were used in the construction of the dams.

To put the Middlemarch (Foulden Maar) deposit in context, a 1995 account of the 'Mineral Wealth of New Zealand' noted that production of diatomite 'is sporadic and probably related to local, rather than national demand'. It mentioned marine diatomites of Eocene age at Papakaio near Ōamaru and at Kaiwaka in Northland, and reported diatomites from lake deposits associated with volcanics of late Tertiary and Quaternary age from near Kamō in Northland, Kingsland, Mercer, Lower Kaimai and Whirinaki in the North Island, and Middlemarch in the South Island.[9] The last reported production was from Whirinaki (1113 tonnes in 1977 for use in hydroelectric dams on the Waikato River) and Middlemarch (564 tonnes in 1966). In 2007, the estimated size of New Zealand's diatomite resources was given as about 5 million tonnes at Middlemarch, 2 million tonnes at Whirinaki, 200,000 tonnes at Lower Kaimai and several million tonnes around Ōamaru.[10]

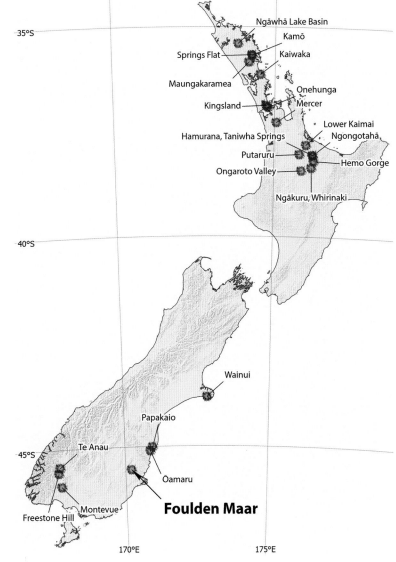

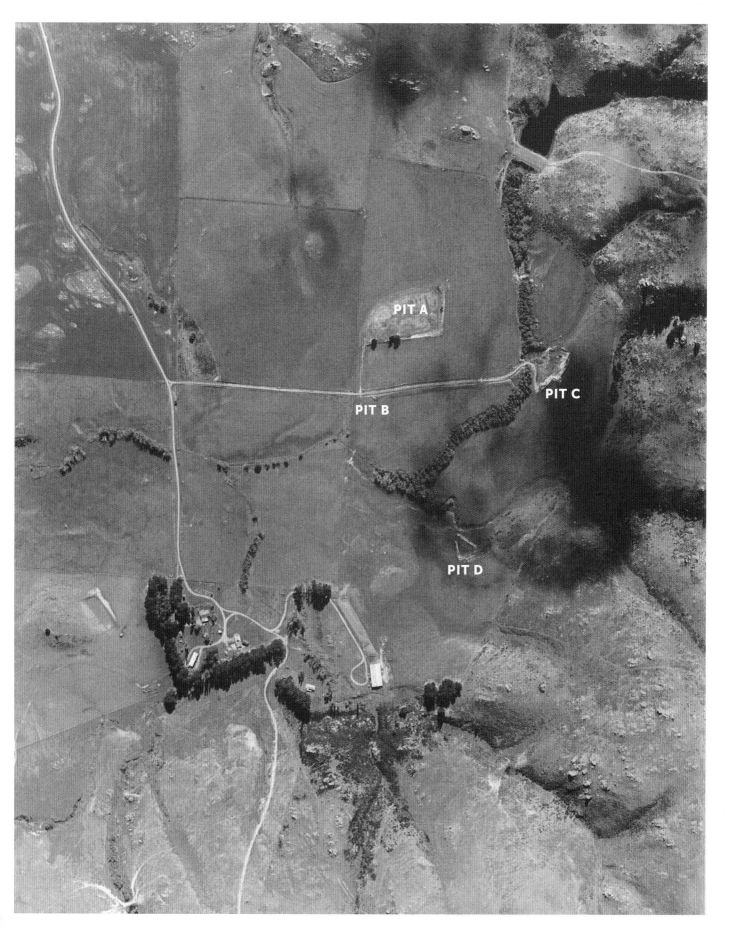

ABOVE The main pit (Pit A) in 2011 with the smaller Pit B in the background. The farm dam in the foreground was built using spoil from the diatomite mining operations. JGC

THE HISTORY OF MINING AT FOULDEN MAAR

Mining records from the 1940s to the 1970s show that a few tons of diatomite were taken from Foulden Maar in most years. In 1941 and 1942, for instance, 79 tons and 74 tons respectively were sold for use in insulation.

About 1967 part of the land surrounding the maar, including the 42 hectares with several small open-cast mining pits, was sold to Southland Cement, which continued mining on a small scale. The company was later taken over by Milburn Cement Holdings, which subsequently became Holcim Cement.

There was renewed interest in diatomite as

an economic resource in the late 1990s. In 1997 Featherston Resources took over the diatomite mine, carried out research and development and in 2011 built a million-dollar processing plant on the Taieri Plain near Mosgiel. The diatomite was destined for use in filtration products and later as a fertiliser additive.

The diatomite deposit was estimated to be more than 35m thick with a reserve of about 5 million tonnes. Dunedin City Council gave consent for mining activity in July 2000 and annual production of up to 25,000 tonnes of dried diatomite was expected to begin in 2001. In the end, however, finding a market proved difficult and very little was removed or processed.

OPPOSITE An aerial photograph of Foulden Maar in 2000 showing the existing mining pits.
Robert Batt and Phillip Marshall (Dunedin City Council)

Featherston Resources was placed in receivership in 2014 and the mining operation was sold to Plaman Global, a company with a reputation for acquiring 'distressed assets'. In late 2015 Malaysian-based Iris Corporation, the majority shareholder in an associate company, Plaman Resources, took over the assets of Featherston Resources. Iris Corporation wanted the diatomite to sell as a fertiliser for palm oil plantations in Malaysia and Indonesia.[11] At that point reserves were estimated at 50–100 million tonnes. This new and much greater figure was likely based on geophysical research carried out by the University of Otago. In November 2017 the Overseas Investment Office (OIO) gave consent for Plaman Global to purchase the 42ha site. The existing mining permit was transferred to Plaman Resources in early 2018.[12]

Reports began to appear in local media and mining journals, indicating that the company was planning a large-scale mine that would destroy most, if not all of the maar – and the fossils and climate records contained within it. An article on mining 'Black Pearl', as the dark-coloured diatomite was henceforth to be trademarked, stated that up to 31 million tonnes would be removed over a 27-year period, this time to be processed into a stockfeed additive for poultry, pigs and cattle.[13]

The trademark was apparently intended to differentiate the dark, wet diatomite from below the water table from the white, weathered dry diatomite on the surface. The dark colour comes from the organic material within the diatomite – mostly fine detritus and plant material such as leaves. When dried out, however, this 'black' diatomite (which contains a high percentage of water) becomes light-coloured 'white' diatomite, identical except in colour and water content. How the company would maintain the black appearance was a mystery.

The company made a further application to the OIO to acquire the surrounding farm so that mining operations could extend beyond the 42ha. It also proposed that the scale of mining be expanded to produce 500,000 tonnes of diatomite each year from 2020 until 2047. At the conclusion of operations, the company claimed, it would leave behind a lake 'for the use of the community'. How it intended to detoxify this 'lake', given that diatomite contains appreciable amounts of pyrite, was not mentioned. Pyrite is involved in acid mine drainage problems: when exposed to water and oxygen, it can react to form sulfuric acid and dissolved iron.

Plaman highlighted the regional employment opportunities associated with expanded diatomite mining, and media reports spoke of more than 100 jobs for local workers in an operation that would run 24 hours a day, seven days a week.[14] Little mention was made of the health hazards associated with the mine, such as working with fine silica dust (crushed diatomite particles). Would the stockpiles be kept wet to reduce the risk of inhalation? The only available water body in the arid environment around Foulden Maar is a small farm dam; the next-closest source is the Taieri River, some 6km away.

Purpose-built 37-tonne truck-and-trailer units would transport the wet diatomite to Milton, 98km away, where a new plant, to be built with $30 million of taxpayers' money from the Provincial Growth Fund, would crush, dry and granulate the diatomite.[15] Other reports mentioned the possibility of enlarging the tunnels on the Middlemarch–Mosgiel railway line to allow trains to transport the diatomite to the processing plant.

Regulations prevent diatomite from being sold as a stockfood additive in New Zealand, so after processing, the dried diatomite was to be shipped to China, Brazil, Indonesia, the Philippines, Vietnam and Thailand.[16]

As more information was released and the New Zealand and international paleontological communities became aware of the scale of the project, community and scientific opposition to the proposal began to build. On 20 April 2019 a newspaper article based on a leaked report revealed much more of what was in store for Foulden Maar. The confidential report, authored by investment bankers Goldman Sachs for would-be investors in Plaman, had fallen into the

View of Pit A in early 2019 showing piles of loose diatomite.

The north wall of Pit A, from a similar vantage point. Note how the diatomite beds dip towards the centre of the maar. An earlier pit (Pit C), now overgrown, can be seen in the background. DEL, JHR

hands of *Otago Daily Times* reporter Simon Hartley.

According to Hartley, the mining proposal appeared to have stalled. The confidential report recommended the company 'remove any link between the Foulden Hills [sic] project and the potential use of Black Pearl as a fertilizer for palm oil plantations', because the topic had 'been identified as a sensitivity for the local community'. Goldman Sachs was confident, however, that the mining company could overcome any resistance encountered: 'Any appeal to the Environment Court is likely to come from a small number of local residents, who are not well-resourced and will not have comprehensive technical reports to the same extent as Plaman Global.'[17]

As the scale of the proposed operation become public knowledge, local residents and the New Zealand and international scientific community together began lobbying to prevent the mining of Foulden Maar. New Zealand media was soon full of feature articles and opinion pieces, and Daphne Lee gave public presentations on Foulden Maar to more than 20 groups, from Invercargill to Auckland.

Online magazine and news site *The Spinoff* invited Juliet Gerrard, the prime minister's chief science adviser, to provide an independent scientific view of the issues surrounding Foulden Maar. Gerrard wrote that in her attempt to discover who might be able to protect the scientific treasures of Foulden Maar, she had fallen into a 'bureaucratic labyrinth'.

On *Newsroom*, under the headline 'Foulden Maar: Unjustifiable vandalism and grand promises', journalist Rod Oram examined the financial aspects of the mining proposal. He noted that the company's US$1 billion annual revenues would work out at US$10 million per employee. 'This would be fabulously profitable for the investors, while the locals would earn only typically modest mining, trucking and processing wages.'[18]

Other articles questioned the purported benefits of diatomite as an animal health supplement. Farah Hancock wrote for *Newsroom*, 'It's not just New Zealand scientists who have doubts about the product.' She noted that North Carolina State University adjunct professor Simon Shane questioned the absence of data from the trials with chickens and pigs that Plaman Resources claimed in a blog post had 'outstanding' results, saying, 'There is no published data in the peer-reviewed literature to support a claim that diatomaceous earth can serve as a growth promoter or enhance feed conversion efficiency in monogastric animals.'[19]

Early in the process, Plaman Resources had lobbied the Dunedin City Council (DCC) and Clutha District Council for support. The DCC was initially positive about the mining proposal, but withdrew its support after information on the scientific importance of Foulden Maar and its fossil treasures became more widely known.[20]

Plaman had stated that the operation would be financially viable only if it could buy the surrounding farmland. This required OIO approval, however, and the weight of opinion in submissions was heavily against. Scientists working at the site joined private individuals and scientific bodies such as the New Zealand Ecological Society and the Geoscience Society of New Zealand to put up compelling arguments for conservation of the site. An online petition called 'Save Foulden Maar' garnered more than 10,000 signatures.

On 13 June 2019 Plaman gave notice that it was being placed in receivership, and six months later the DCC announced it would take over responsibility for the 42ha Foulden Maar site.[21] The immediate threat was gone.

At the time of writing, negotiations about the ownership and future access to the site are ongoing. The long-term protection of the Foulden Maar diatomite deposits is not yet assured.[22]

FOULDEN MAAR – A BUREAUCRATIC LABYRINTH?

I've been lobbied to take up the cause and shout 'SCIENCE!' at the government, thereby joining the chorus of over 5,000 who have signed the petition – 'We are calling on you to stop the destruction of Foulden Maar, near Middlemarch in Central Otago, by Plaman Resources Limited'. This sort of suggests that someone, somewhere, is sitting in an office making a clean choice between fossils and profit – so I set about finding out where this office might be. This has proved to be something of an archaeological dig in its own right. There is a hint that this might be the case in that said petition is actually addressed to the prime minister, nine ministers and two councils. Undeterred, I called the chief science advisors for the ministries who support the ministers addressed by the petition to see who might know the answer to my question of who could have dismissed the scientific case and allowed the mining to proceed.

[...] Roughly speaking, these decision-making processes fall outside of the stewardship of: the Ministry for the Environment and the Environmental Protection Agency (unlike, say, a river); the Department of Conservation (unlike, say, a kākāpō nest); and Heritage New Zealand (unlike, say, an historic building).

Mining permits are granted through MBIE, who look at the technical and financial aspects of the proposal, and meeting environmental legislation (which there may be for rivers, kākāpō nests, and historic buildings, but seemingly not for fossils). So the scientific case features only in local government processes – in this case the Dunedin City Council and the Otago Regional Council – under the parallel resource consenting processes that are required before you can start mining with your permit.

The science case, then, needs to be made at local government level, at a hearing around resource consent, if there is a hearing. There is probably another archaeological dig through council processes here ... The key point here is that there should surely be a point in central government where national scientific treasures could be considered as a factor in the permitting process?

What about that purchase of the neighbouring land? Because the proposal to purchase this block is by an overseas purchaser, it has to go through the Overseas Investment Office. This office is not in MBIE but in LINZ (and so the list of ministers grows). There is a thin ray of hope here that central government could rule in favour of the fossils, but that would require 'special scientific status' (for example, if it were a UNESCO site).

Incidentally, were the private owners to want to apply for designation of the area from UNESCO, an application would be made through UNESCO-NZ which is under the Ministry of Education. Indeed, a Geopark has been proposed in nearby Waitaki, but the application has been delayed. Foulden Maar is not within this proposed site, and so any special designation of the site is a subjective decision. In summary – it is very hard to find where the rallying cry of the scientists can be simply heard.

— Juliet Gerrard, *The Spinoff*, 20 May 2019

CHAPTER 3

LIFE IN THE LAKE

A deposit of diatomaceous earth on the Strath Taieri … is chiefly made up of three or four minute species of Cymbella or Cocconema … although I have not been able to determine any of them, I have no doubt but that they are freshwater species.

– Hutton & Ulrich, 1875[1]

New Zealand today is a land of lakes. Ranging in area from a few hectares to more than 600km^2, they have been formed by processes as varied as glaciation, landslides and volcanic activity. All lakes are ephemeral – short lived – on a geological time scale. Sediment carried in by streams settles on the lake floor and eventually fills it in. These lake sediments often hold extremely detailed records of geological events, climate and the biota that lived in and around the lake.

A paleolake is one that existed in the distant past, and Otago has many of these. Most are of Miocene age. The largest was Lake Manuherikia, which occupied an area of 5600km^2, greater than all extant lakes in New Zealand combined. One of the smallest was Foulden Maar, which, although only 1000m in diameter, holds a remarkable archive of the plants and animals that made up the lake ecosystem throughout its entire 130,000-year history. From microscopic diatoms and other algae, freshwater sponges and insects to galaxiid fish and large eels, virtually all the species that once inhabited the lake were 'pickled', layer by layer, on the lakebed. These wonderful fossils are now gradually being revealed as researchers carefully tease the layers of sediment apart.

DIATOMS

Most of the sediment in the Foulden Maar crater consists of billions of diatoms. These microscopic, single-celled aquatic algae live in a tiny opaline silica 'box' called a frustule, which may be preserved in the fossil record.

BILLIONS OF TINY SILICA BOXES

Diatoms are key components of the phytoplankton, the microscopic algae that form the base of the food chain in many aquatic ecosystems. They have a long fossil history extending back to the Cretaceous. Their frustules range in size from 1μm (1 micron) to 1mm, and they are present in most water bodies, from freshwater to brackish and marine. They can be solitary or colonial (colony-forming), and they can be planktic (living near the water surface) or benthic (bottom-dwelling). Some live in soil, others in ice. There are thousands of diatom species and more than 300 living and fossil genera; about 70 percent of these are exclusively marine. Diatoms are most abundant in seas that have high levels of nutrients, and they make up about 40 percent of the organic matter found in oceans.[2]

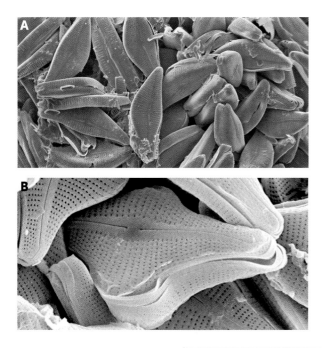

Classified by their shape, diatoms may be centric (wheel-like) or pennate (oblong or boat-shaped). Most of those at Foulden are pennate and belong to a single species, *Encyonema jordaniforme*, known only from this locality.[3]

This diatom species must have arrived in the lake soon after the eruption and literally took over. Every spring and summer for over 130,000 years they bloomed in unimaginable numbers in the surface waters, and every autumn and winter the dead diatoms sank to build up a thin layer on the lake floor. It is likely

A|B Two scanning electron microscope (SEM) images of the pennate *Encyonema jordaniforme* diatoms showing fine ornamentation on the valves. **C** A light microscope image. UK, EG, JMB

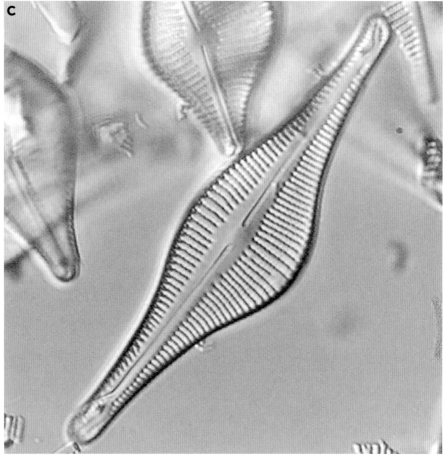

that they were mucilaginous – bonded together with a gum-like substance – which also helped preserve the rest of the biota that sank to the lakebed.

It took almost 150 years to establish the correct name for the main diatom species at Foulden Maar. In 1875 Hutton and Ulrich listed 'three or four minute species of *Cymbella* or *Cocconema*'.[4] Later researchers applied various names such as *Cymbella minuta* and *C. jordanii*, and these were later transferred to the genus *Encyonema*. Until 2018 the diatoms were generally referred to as *Encyonema jordanii* (Grunow ex Cleve 1894) Mills 1934. However, taxonomic detective work by Margaret Harper and colleagues in Belgium, Denmark, Austria and New Zealand established that the species at Foulden did not match the type and other specimens of *Cymbella jordanii* held in museum collections in Europe.[5] Mystery surrounds the 1894 specimens described originally by Cleve – who collected them, and from where? None have been collected from anywhere in New Zealand either before or since, and the manuscript in which they were described is now lost.[6]

So what *was* the correct name for these diatoms? Another European diatom researcher, J. Foged, visited New Zealand in 1966, and some of the material he collected was later described and illustrated by the diatom taxonomist K. Krammer as a new species, *Encyonema jordaniforme*. The type locality for the species – the place where it was first identified – is listed as Te Anau, but a check of the actual specimens held in the Foged Collection at the Statens Naturhistoriske Museum in Copenhagen revealed that they are labelled 'Middlemarch district', which is about 180km from Te Anau. This almost certainly refers to Foulden Maar, as the freshwater diatomite here is dominated by this species. *Encyonema jordaniforme* Krammer 1997 thus became the accepted name for the dominant Foulden Maar diatoms.[7]

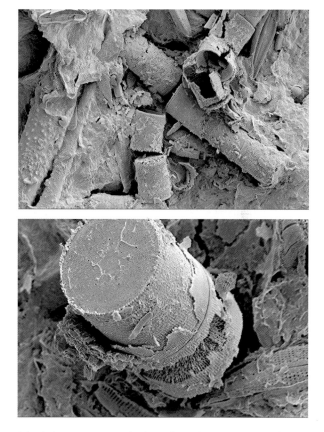

Much less common in the winter layers are centric diatoms. These fossils from the spiculite bed resemble *Aulacoseira*.[8] UK

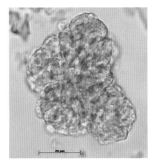

LEFT *Botryococcus* prefers quiet, slow-moving water with a low sediment influx. DCM

GREEN AND GOLDEN-BROWN ALGAE

Species of algae without silica cell walls also made up part of the phytoplankton in the maar lake. The colonial green algae *Botryococcus*, which forms blooms and sometimes large floating mats on modern lakes, is found sporadically in pollen samples in Foulden. In marked contrast to the local diatom species, which is found nowhere else in the world, *Botryococcus braunii* is widely distributed worldwide in freshwater and lakes from temperate to tropical environments. Because it can synthesise long-chain hydrocarbons, lake deposits rich in this alga sometimes fossilise to become oil-bearing deposits, such as the Nevis Oil Shale in the Nevis Valley in Central Otago.

Other intriguing components in the dark layers of the diatomite are stomatocysts of the golden-brown microalgae Chrysophyceae.[9] The soft parts are tiny free-swimming single cells just 5–10μm in diameter, and these leave no trace. However, the resting spores or stomatocysts – cells that survive over winter – are tiny hollow silica balls with a single aperture, and these do fossilise. This opening has an organic or siliceous plug or lid that is sometimes also preserved. Although little studied, they are common in lake sediments and prefer low levels of nutrients.

LEFT Close-ups of two of the minute, hollow globular resting spores that survived in the Foulden lake over winter. JKL

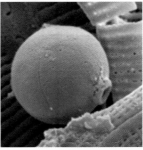

A Laminated diatomite at Foulden Maar, showing the annual layering of the sediment. **B** An artificially coloured SEM image of a cross-section of laminated diatomite showing a denser (winter) layer containing sponge spicules (green), centric diatoms (yellow) and chrysophycean stomatocysts (red), sandwiched between two spring/summer layers full of silica frustules from pennate diatoms. UK

SPONGES

Also scattered among the myriad diatoms are the glassy spicular (spike-like) remains of some animal inhabitants of the lake: freshwater sponges.

Sponges are among the simplest of multicellular animals, with a tubular structure that allows water carrying food and oxygen to flow through their tissues. Sometimes described as 'primitive' because of their simple body plan, sponges have thrived on Earth for hundreds of millions of years. Although better known as inhabitants of the marine realm, a few sponge species live in lakes and rivers: New Zealand has 724 marine species and only five that live in freshwater. Freshwater sponges are quite common, yet little research has been carried out on them.[10] The living New Zealand freshwater species are placed in the family Spongillidae as members of the genera *Ephydatia*, *Eunapius*, *Heterorotula* and *Radiospongilla*.[11]

Sponges are mostly soft-bodied animals consisting of undifferentiated tissue, so the only parts that fossilise are the siliceous spicules that once held the sponge together. The spicules vary in shape and occur in three types: megascleres – long, tapered spicules that form a mesh/network to support the tissue; microscleres, which are smaller; and tiny gemmuloscleres, which

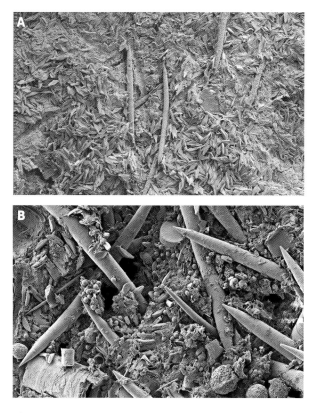

A|B SEM images of elongate glassy sponge spicules resting on a bedding plane of much smaller diatoms. JMB, UK

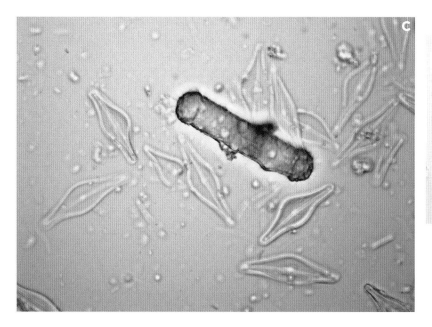

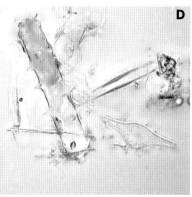

C|D Light microscope images of sponge spicules among diatoms. Spicules are common throughout the diatomite. JMB

are found inside small cell masses called gemmules – small resting spores produced by freshwater sponges when the environment becomes unsuitable, which will bud into a new sponge when conditions improve. The sponge fossils found at Foulden are assigned to the genus *Spongilla* and include examples of all three spicule types.

Isolated spicules are found throughout the deposit, but a 60cm-thick bed exposed in the upper part of the diatomite in Pit A is so packed with sponge spicules that the sediment is described as a 'spiculite'. Here, sponges replaced diatoms as the major component of the sediments for an interval of about 1000 years. Interestingly, the diatoms present in this sponge layer are mostly benthic or bottom-dwelling centric diatoms, in contrast to the pennate *Encyonema* that were the overwhelmingly dominant planktic component of the lake biota for over 100,000 years. This change might have been due to increased turbidity in the lake, changes in water chemistry and/or fluctuating water levels.

The sponges provided microhabitats for other organisms such as bacteria, algae and insects and their aquatic larvae, which were also a crucial part of the lake ecosystem (see Chapter 6). Sponge-dominated mats may have floated on the open water surface, as well as colonising the sediment surface and water plants.

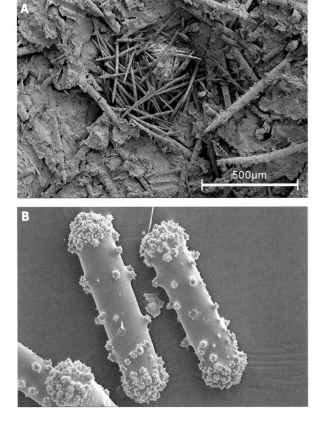

A Spiny sponge gemmuloscleres surrounded by slender megascleres form a 'nest' on a bedding plane. **B** Two short stout gemmuloscleres. Note the exquisitely preserved minute tubercles on the tips (less than 1μm). UK

OPPOSITE In the north wall of Pit A (see bottom photograph, page 45), a 60cm-thick bed packed with sponge spicules is sandwiched between typical layers of diatomite. UK

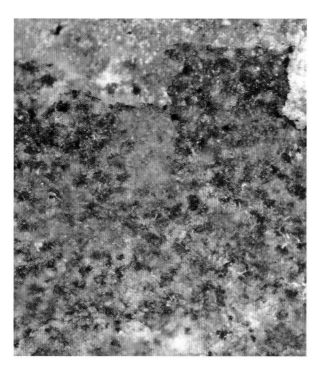

FRESHWATER FISH – THE WORLD'S EARLIEST WHITEBAIT

Macrofossils are typically more than 1mm across and visible to the naked eye. The first to be found at Foulden Maar were leaves and fish. Two small, articulated but poorly preserved fish were collected by Cecilia Travis in 1962. She sent photographs to New Zealand freshwater fish expert Gerald Stokell, who identified them then as *Galaxias kaikorai*, a fossil species named a few years earlier from Frasers Gully in Kaikorai Valley, Dunedin.[12]

Galaxiids belong to a distinctive Southern Hemisphere group of freshwater fish that lack scales, have no adipose (posterior dorsal) fin and whose dorsal fin sits above the anal fin. About 21 species are currently recognised in the modern New Zealand fauna. Several, including the five species whose juveniles are known as whitebait, were once abundant

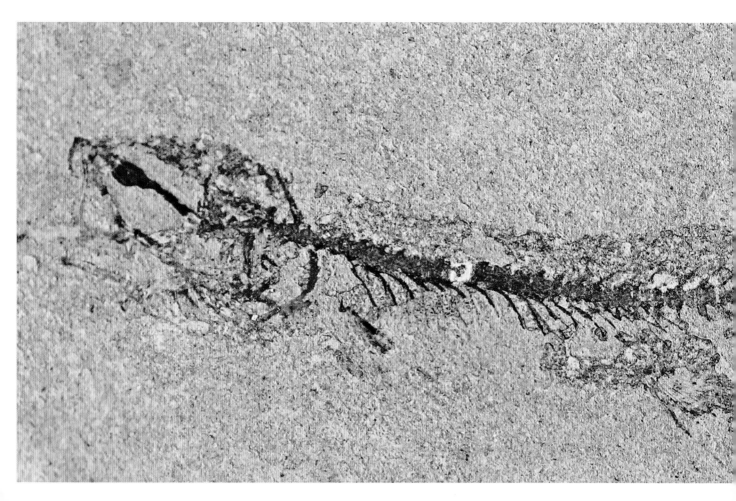

in many New Zealand rivers but are now under threat from habitat modification and overfishing.

Galaxiids in New Zealand are variously known as galaxias, mudfish, īnanga, īnaka and kōkopu, among many other Māori names for fish at varying life stages. The family name Galaxiidae references the silvery-gold star-like patterning on many species.[13] Long considered to be of Gondwanan origin, galaxiids are also found today in Australia, New Caledonia, South America and the Falkland Islands. Charles Darwin speculated in 1872 on possible explanations for this wide Southern Hemisphere distribution.[14]

In New Zealand today, galaxiid species occur in streams, rivers, lakes and swamps from Northland to Stewart Island and as far afield as the Chatham Islands and the Campbell Islands. Most species are diadromous, meaning they spend the early part of their life at sea before returning to a freshwater environment

BELOW Entire galaxiid specimens with gaping mouths. Some have brown to blackish 'shadows' that are remnants of the soft tissue (skin and eyes) of the living fish. The remarkably preserved skin patch (**OPPOSITE**) has been photographed under ethanol. UK

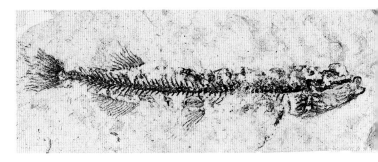

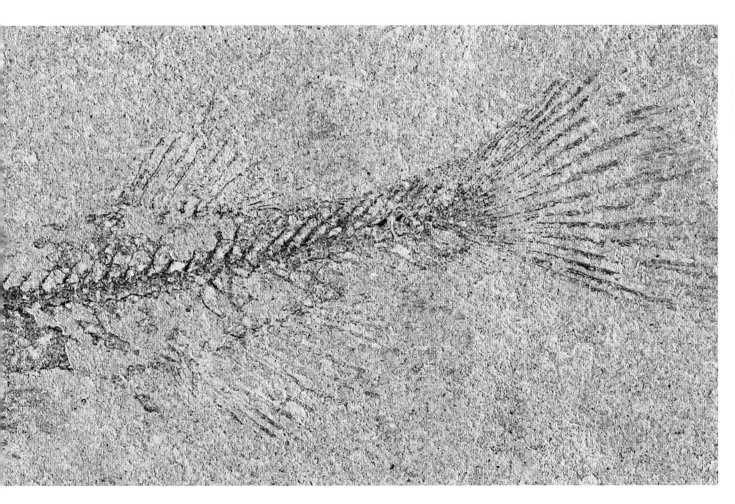

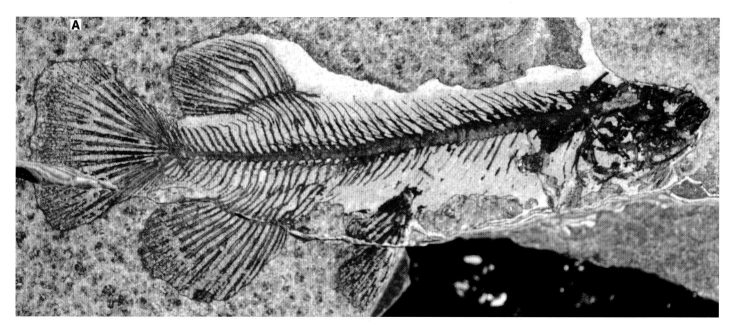

A

to reproduce. A few species are land-locked and non-diadromous (from choice or chance) and their entire life cycle takes place in lakes.

The only known galaxiid fossils in the world are in southern New Zealand. Those found at Foulden Maar are 23 million years old, the oldest so far.

Articulated galaxiid fossils of Miocene age found in other Otago paleolakes include the *G. kaikorai* from Frasers Gully (12 Ma), a new species from a diatomite deposit at Kilmog Hill north of Dunedin (11 Ma), and another new species, as yet undescribed, from the carbonaceous mudstones in the Hindon Maar Complex between Outram and Middlemarch (14 Ma).[15] Galaxiid skull components that include jaws and teeth, and one large headless specimen, have also been found near Bannockburn in Central Otago, and over 200 otoliths (fish earbones) provide evidence that six or seven more species once swam in the vast early to mid-Miocene paleolake Manuherikia, also in Central Otago (19–16 Ma).[16]

The best-preserved and most complete fish to be found at Foulden was effectively 'filleted' when the column of laminated diatomite in which it occurred was split with a field knife; one side was preserved on the top layer and its counterpart on the bottom. The fish is stout-bodied compared to most living *Galaxias*, and differs from all modern species in its relatively low vertebral count and expansive dorsal, anal and caudal or tail fins.[17]

These small fish must have been common in Foulden Maar lake, as more than 100 galaxiid fossils have now been collected from the two main mining pits. Most are articulated, laterally compressed skeletons and range from 40mm larvae (whitebait) to 140mm-long adults. Randomly dispersed throughout the exposed diatomite sequence, the little fish show no preferred orientation, and no mass mortality layers have been found. The exceptional preservation of soft tissue – eyes, skin and even skin patterning – is one of the reasons why Foulden Maar qualifies as a *Konservat-Lagerstätte* deposit.[18]

The stratified nature of the lake meant that only the upper few metres had sufficient oxygen for fish to thrive. The deeper waters were anoxic, and the Foulden fossils may have succumbed to low oxygen from swimming down too far (many specimens died with gaping mouths), or else they sank after dying from other causes in the upper layers. In the absence of

A|B The 14cm fish collected by Daphne Lee in 2005, which became the holotype of *Galaxias effusus*. The species name refers to the 'lavish or extravagant' fins. The original thin, delicate fish bones are generally dissolved (decalcified) and are preserved as soft brown to black residual material. **C** Close-up of some of the delicate bones that have been partially replaced or coated by silvery-grey pyrite, the iron sulfide mineral also known as 'fool's gold' that is common in the deposit. JKL

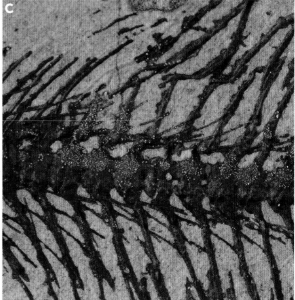

scavengers at depth they lay undisturbed, to be covered by successive seasonal layers of diatoms.

There are a couple of ways in which *Galaxias* may have entered the isolated maar lake in the first place. They may have been transported by waterbirds, or juvenile galaxiids may have 'climbed' into the lake from nearby streams, perhaps when the lake overflowed. Some living galaxiids, including the kōaro (*G. brevipinnis*), are strong climbers, able to migrate upstream past steep waterfalls using their broad pectoral fins and surface tension to negotiate wet rocks.[19]

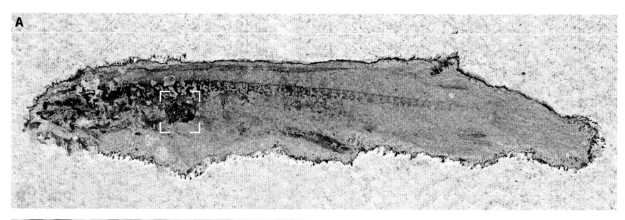

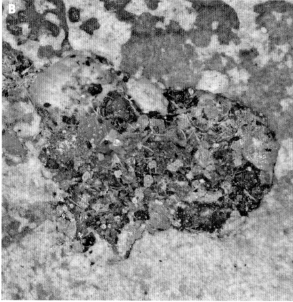

A|B The last meal of this small galaxiid is still visible. UK

COPROLITES

Living galaxiids in New Zealand are generalised invertebrate predators, dining on many species of beetles, moths and caterpillars as well as aquatic insects, such as caddisflies, mayflies, midges and their larvae.

The remarkable degree of preservation at Foulden Maar allows us to identify the food of their 23-million-year-old relations, because indirect evidence of their final meals is visible in the form of coprolites. These fecal pellets have provided important clues to the diet of many long-extinct animals, including dinosaurs, birds and fish.[20]

Thousands of small coprolites have been found at Foulden. These are divided into 10 types on the basis of shape, size and content, and at least two can be assigned to *Galaxias*. The more common of these two is 3–30mm long, brown to black, and may be ovoid, elongate or almost round (subrounded). They consist of leaf and other plant debris, sponge spicules, mineral grains, and the remains of terrestrial and aquatic insects and spiders. The composition of the coprolites is identical to the stomach contents found in some galaxiid fish.

The second type of galaxiid coprolites are 2–9mm-long whitish strings that consist almost entirely of pennate diatoms and some organic detritus. These were produced by *Galaxias* larvae, which live exclusively on algae such as diatoms.

C A *Galaxias* larva showing the characteristic large eyes. UK

Two different coprolite types can be attributed to other lake inhabitants. The first is larger (up to 32mm long) and made almost entirely of polished grains of fine quartz sand. No such sand grains are present within the maar lake basin sediments, so these must have originated in streams beyond the maar. We speculate that waterfowl ingested sand with their food while foraging in shallow water nearby, before flying on to the Foulden paleolake.

The second type of coprolite is rare: it includes the bones of galaxiid fish and provides indirect evidence of a larger predator or scavenger in the lake.[21]

EELS – THE TOP LAKE PREDATOR

The predators at the top of the food chain in the maar lake were freshwater eels (Anguillidae).

Today eels are widespread, although threatened, in rivers, lakes and swamps throughout New Zealand. Until now it was unknown how long they had been part of the freshwater fish fauna. There were no records of fossil freshwater eels from anywhere in the Southern Hemisphere until one was discovered at Foulden Maar.

Anguilla species are distinguished by their elongate tubular body form, numerous vertebrae, a small pointed head, vomerine teeth (small projections for securing prey) in the roof of the mouth just behind the

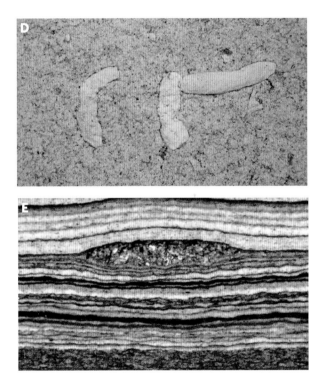

D A bedding plane with white diatom-rich coprolites produced by galaxiid larvae. **E** A cross-section of a waterfowl coprolite made up largely of polished quartz grains. UK, JKL

upper jaw, and dorsal and anal fins that are continuous with the tail. The evidence suggests that the fossil eels from Foulden (and other better-preserved specimens from Hindon Maar) are closest to the endemic (native to and found only in New Zealand) longfin eel *Anguilla dieffenbachii*.

In carrying out this identification we considered alternative candidates from the Australasian freshwater fauna. The presence of fins excluded terrestrial taxa such as snakes and their allies. The only other extant freshwater fish with an elongate shape in the New Zealand biota is the lamprey, *Geotria australis*. Lampreys lack jaws, however, and a jaw structure is clearly evident in the fossil. The Australian freshwater fish fauna contains several swamp eels (synbranchids), but these have an overshot top jaw, no pectoral fins and very rudimentary short dorsal and anal fins. These features do not match the fossil.[22] The evidence is strong for a species similar to *A. dieffenbachii*.

Eels, like many *Galaxias*, spend part of their life cycle in freshwater and part at sea. However, while galaxiids return to freshwater only to lay eggs and breed, eels live for most of their often very long lives in freshwater. They must then return to a particular deep-water region in the tropical ocean to breed, a life habit known as catadromy.

New Zealand today has two native eel species: the longfin *Anguilla dieffenbachii* and the shortfin *A. australis* (both known to Māori as tuna). A third species, *A. reinhardtii* (spotted eel), has extended its range in the last few decades from the east coast of Australia and New Caledonia to northern New Zealand.[23]

Longfin eels occur throughout New Zealand, including Rakiura Stewart Island and Rēkohu

Chatham Islands. They live in wetlands, streams, rivers and lakes from sea level up to 1150m, and are sometimes found offshore at depths of more than 300m. Females of *A. dieffenbachii* may reach 2m in length and weigh as much as 25kg; males are smaller. They mature at about 35 years but may live on in freshwater until they are more than 80 years old. In their final autumn they become desperate to return to the sea and will migrate several thousands of kilometres to the subtropical Pacific near Tonga, where they spawn millions of eggs. Despite many attempts to track adult eels on this migration, the exact location of this spawning ground remains uncertain.

The smaller shortfin eel grows to 1.2m and reaches only 3.5kg in weight. Females mature at 30 years and males at 15; and they also migrate on their final journey to the western subtropical Pacific.

About 18 months after hatching, the leaf-like larvae or leptocephalae of both species journey to New Zealand, where they enter the rivers as 60–75mm-long transparent glass eels before turning brown and beginning the freshwater part of their amazing life cycle. Small eels live on insects, but by the time they reach about 80cm their diet may also include fish and small birds.

Young eels (less than 120mm long), like galaxiids, are skilled climbers and can ascend high, steep waterfalls and travel long distances over wet ground. They may have found their way into the maar lake overland from nearby bodies of water, possibly after heavy rain, or even through subsurface aquifers. Unlike the *Galaxias*, however, which could complete its life cycle within the little lake, the eels in Foulden Maar were probably trapped, surviving perhaps for decades but unable to return to the sea to breed, as sometimes happens to eels trapped in lakes behind dams in New Zealand today.[24]

In 2010 Uwe Kaulfuss found a block of diatomite that appeared to contain parts of a slender elongated fish. Over time he found other parts in further blocks, and eventually he fitted the jigsaw together. The decalcified vertebral column preserves at least 54 vertebrae, while the narrow, continuous dorsal and anal fins appear as brownish smudges. The conical, slightly curved teeth at the anterior end of the jaw are arranged in dense rows. UK

CHAPTER 4

FOSSIL PLANTS 1: FERNS, CONIFERS AND FLOWERING PLANTS

To reconstruct the forests of the distant past is a challenging undertaking. At most fossil sites in New Zealand, the only evidence of the trees and shrubs that once formed the vegetation cover comes from pollen and spores, and these minute microfossils are often difficult to interpret, as they may have been transported tens to hundreds of kilometres from their forests of origin.

At Foulden Maar, however, it is possible to recreate the rainforests that surrounded the lake in exquisite detail. All the myriad leaves, flowers, fruits, seeds, wood and bark found in the lake sediments are from trees, shrubs, climbers, epiphytes and smaller plants that grew around its margin. They were either blown or fell into the water, eventually drifting down to the anoxic lakebed where they were preserved with their fine cellular details intact.

Plant remains are by far the most common fossils at Foulden. The sediment itself is mostly made up of microscopic algae, but these can only be seen with a high-powered microscope. Anyone visiting the maar armed with a collecting knife or geological hammer will, within the first five minutes, split a layer of diatomite and discover a leaf that has not seen the sun for 23 million years. Occasionally, on a windy day, a leaf will lift off the surface in the breeze and will need to be recaptured.

Each individual leaf tells a story about its parent plant. At Foulden the plant fossils also tell complex stories about ecological interactions. Flowers with pollen preserved in their anthers give clues about pollination by wind or insects; fruits and seeds hint at dispersal by birds. Scale insects have been found, still attached to the leaf from which they were sucking sap when a gust of wind carried them both into the lake.

And the story carries on for more than 100,000 years after the eruption devastated the landscape 23 million years ago. It can be found in the charred pollen at the bottom of the core, in the evidence of revegetation by ferns and scrub, and in myriad examples that tell of the eventual establishment of a highly diverse subtropical rainforest on the fertile volcanic soil. Few other sites on Earth allow us to reconstruct an ancient terrestrial ecosystem in such detail.

OPPOSITE From left: Liz Kennedy, Dallas Mildenhall and Uwe Kaulfuss preparing to collect pollen samples at Foulden Maar. DEL

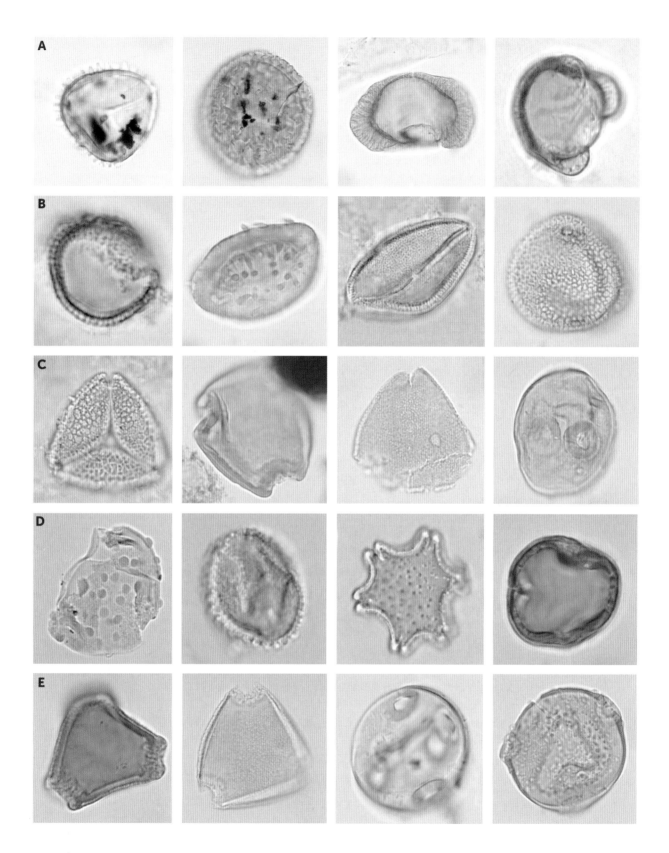

POLLEN AND SPORES – MICROSCOPIC SPLENDOUR AND DIVERSITY

Palynology is the study of pollen and spores from both living and fossil plants. Flowering plants shed vast numbers of pollen grains that are transported by insects or wind to other flowers in the process of pollination. If carried by insects, they may be moved only a few metres from the parent plant, but if they are windborne they may travel many kilometres. During pollination, the male DNA in the pollen grain reaches the female ovule in another flower, and the resulting fertilisation produces mature seeds. Spores are equally small reproductive structures, and are produced by ferns and fungi.

Although tiny, pollen and spores are very resistant to decay and can be preserved for millions of years in fine-grained sedimentary rocks.

The outer coats of pollen and spores often carry distinctive ornamentation that makes them easily recognisable. Even in the absence of plant macrofossils, by studying fossil pollen and spores extracted from sediment samples, scientists are able to reconstruct plant communities and paleoclimates of the past.

Small pollen samples were taken from the base to the top of the Foulden core, and others from the sections exposed in the pits. Examination of these under a high-powered microscope in the GNS Science Paleobotany Laboratory in Lower Hutt revealed thousands of pollen grains: 300 were counted from each sample, making a total of 10,000 individual pollen grains and spores identified from throughout the layers of sediment. These carry an amazing story of changing vegetation in the area from just before the maar-forming eruptions to the end of the lake record, 130,000 years later.[1]

The palynological study of the Foulden core revealed a vast amount of information.[2] The pollen and spores from the bottom section of the core are charred and blackened and charcoal is present – evidence of the violent eruptions that incinerated the existing broadleaf–podocarp rainforest when the maar was formed. Pioneer ferns such as *Gleichenia* (coral

OPPOSITE Examples of some of the tiny spores and pollen grains from Foulden Maar.
ROW A Ferns, fern allies and conifers.
ROW B A basal angiosperm and monocots.
ROWS C—E Various eudicot pollens from a range of families. EMK, DCM

BLOWN FAR FROM HOME

Nothofagus species or southern beech, some of the most familiar forest trees in the modern New Zealand flora, produce large amounts of wind-dispersed pollen that can be blown over vast distances. For example, no southern beech trees grow on Rēkohu Chatham Islands, 800km east of the South Island of New Zealand, but the pollen is common there. It is also common in the New Zealand fossil record, again sometimes far from the source trees. In contrast, plants that are pollinated by insects generally produce small amounts of pollen that are less likely to be found far from the parent plant.

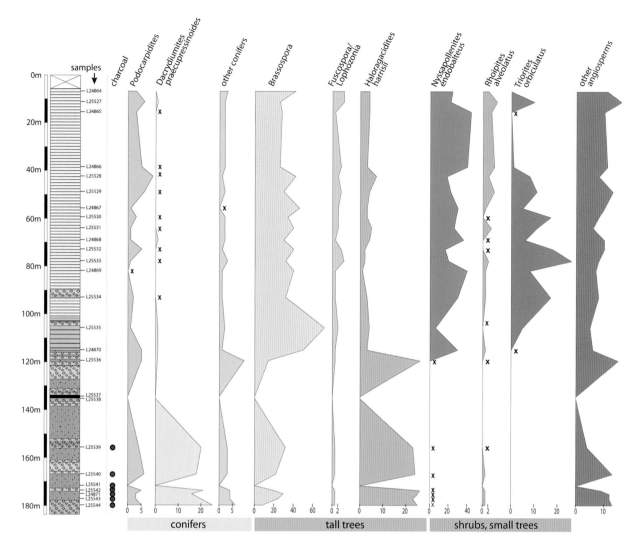

Relative abundance record of selected spores and pollen from Foulden core FH2. The L numbers are the palynology slide collection numbers held at GNS Science, Lower Hutt. The percentages (bottom line) include all spores and pollen (note changes in scale to incorporate low percentages). In the sedimentary column, laminated diatomite (top 106m) is shown in yellow; sections dominated by turbidites (sediments deposited from a low-density mix of sediment and water) (middle 13m) are green; grey indicates graded mudstone and sandstone with volcanic and schist breccia (63m); purple is unsorted schist breccia (5m).

Mildenhall, Kennedy et al., 'Palynology of the early Miocene Foulden Maar', *Review of Palaeobotany and Palynology 204*, 2014, pp. 27–42

fern), early colonisers of bare ground, soon became established on the devastated landscape around the maar.

Within a few centuries the fertile volcanic soil formed on the tephra around the maar had been revegetated by shrubs and small trees such as *Mallotus* and *Macaranga* (spurges or Euphorbiaceae) – many of whose modern species are pioneers of disturbed rainforest margins – and laurels (Lauraceae), which are not seen in the pollen record at Foulden but are the most abundant fossil leaves in the diatomite.[3]

LEAVES

Leaves are by far the most abundant macrofossils and are present in their tens of thousands. Most are entire but isolated, lying flat as they landed on the lakebed after drifting down through the water column. Few of these are still attached to a stem, but many retain attachment petioles (stalks).[4]

Most of the fossilised leaves are of medium size (2–10cm long). A small number are under 2cm, and only a few are very large leaves, though some are 20–30cm. Because the individual layers (laminae) of diatomite are so thin (0.4–1mm), exposing a single leaf on a bedding surface is sometimes difficult, and it can be a challenge to collect a large leaf intact. The absence of tiny leaves could be due to their rarity in the forest or their fragility – probably the former.

Unrelated leaves occasionally land on top of each other in random orientations. DEL

BELOW Jennifer Bannister sorting leaves in the field. DEL

LEAVES IN THE LAB

In the field, a quick check confirms whether the leaf margin is entire (smooth) or serrated (toothed). The type of petiole and the leaf tip are examined if they are present, along with the vein pattern (venation). The diatomite block containing the leaf is then trimmed and sealed in a plastic bag to prevent the specimen drying out.

Back in the lab the leaf block is gently washed with water, and any overlying debris or sediment is removed with a fine paintbrush or needle. The leaf is photographed then soaked in a shallow bath containing hydrogen peroxide until it becomes semi-transparent. After rinsing and further cleaning with a fine paintbrush if necessary, more photographs are taken. Each leaf must be treated individually, as what works for a thick robust leaf may destroy a thin fragile one.

Small cuticle samples (the waxy layer that covers the leaves of all plant species) are cut out of any particularly interesting leaves and mounted on glass microscope slides in glycerine jelly. Staining the cuticles with 0.1 percent crystal violet (a biological stain) enhances the cellular details when viewing with transmitted light microscopy.

Other leaves are dried slowly and stabilised by treatment with the clear resin polyvinyl butyral. If left, they may dry out and disintegrate over a period of days, months or years, depending on the leaf type.

Each leaf, flower or fruit is given a unique field collection number and a corresponding catalogue number in the Geology Museum, Department of Geology, University of Otago. DEL

A|B Most leaves are dark-coloured, often brownish or even black when collected. Occasionally intact leaves will lift off the freshly exposed surface in the wind. **C|D** Some leaves can be lifted off the diatomite bedding planes and mounted whole in glycerine jelly between glass slides. JMB

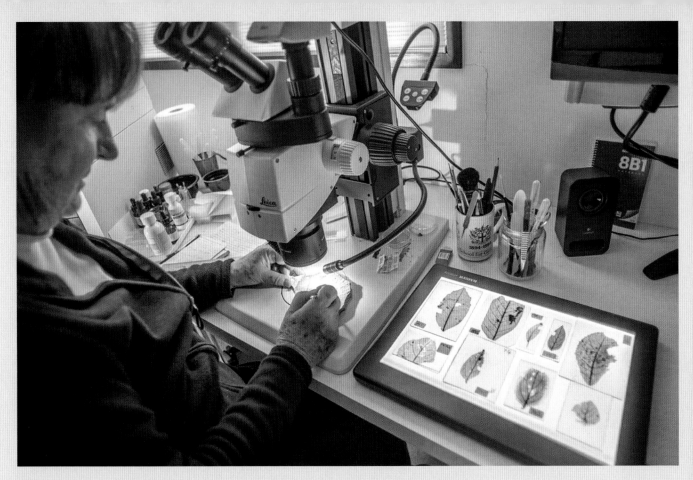

Jennifer Bannister preparing a leaf for photography. To date she has prepared over 800 individual leaves and cuticles. Rod Morris

Cuticles are commonly stained to better show the detail of cell shape and disposition, stomata (the pores in a leaf through which gas exchange takes place) and trichomes (hair bases). The deep purple colour comes from crystal violet, a protein dye sometimes called gentian violet and used as an antiseptic. JMB

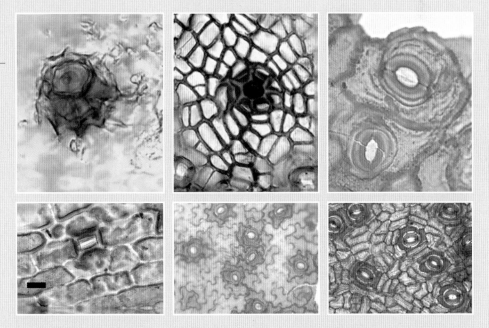

LEAF IDENTIFICATION

Living plants can be easily identified by their stems, bark, leaves, cuticle and especially flowers. A standard protocol has evolved for the identification of isolated fossil leaves, based on leaf architecture features such as overall shape, length and width, type of base, angle of apex, type of margin, venation type, order of veins and other features, such as the presence of domatia – tiny pockets or hair-tufts in the main veins of a leaf, produced by the plant to provide a home for mites.

Well-preserved fossil leaves retain the details of both the upper and lower epidermis of the leaf on the cuticle layer. Details of the cuticle usually allow most fossil leaves to be assigned to a plant family and genus with confidence. Cuticle anatomy varies greatly – as much as the leaves themselves – but is likely to be similar within a given plant family/genus. Some are thick, tough and leathery, such as laurels, whereas others are thin and fragile, such as those of aquatic plants. For example, the cuticle of modern-day species of the deciduous *Fuchsia excorticata* (kōtukutuku, tree fuchsia) is very thin and unlikely to fossilise well, and although flowers, anthers and pollen found at Foulden indicate that a species of *Fuchsia* grew around the lake edge, no leaves have been found.[5]

Once leaf cuticles of the fossils have been prepared, photographed and described, they are compared with reference cuticles of modern leaves from New Zealand, Australia, New Caledonia and elsewhere.

Sometimes it is not the presence but the *absence* of particular leaves that tells the story. The most abundant

pollen at Foulden comes from several extinct species of wind-pollinated *Nothofagus*, but, of the thousands of leaves so far examined, not one southern beech leaf has been found. In contrast, about half the leaves at the mid-Miocene fossil sites of Hindon Maar and Frasers Gully in Kaikorai Valley are *Nothofagus*, and they are common and diverse at a range of other Miocene-aged locations in the region.[6] Since these leaves are robust, distinctive and preserve well, their absence in the Foulden flora indicates they were not growing close to the lake. The pollen has blown in from afar.

The abundance of laurel leaves (40 percent of all leaves collected so far, representing at least 10 species in three genera) is indicative of a Lauraceae-dominated rainforest around Foulden Maar.[7] On the other hand, no pollen of this family is preserved at all, so a solely pollen-based reconstruction of the Foulden forest would be totally misleading. Evidence from all plant remains must be integrated in order to gain a reliable picture.[8]

These leaves from Foulden Maar showcase some of the wide variety of leaf shape, complexity, margin and venation types present at the site.
JMB, JGC

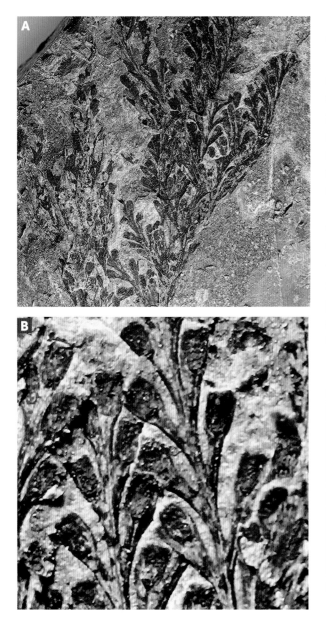

FERNS

There have been some exciting fern finds at Foulden.

Ferns and their allies (pteridophytes) are distinctive components of New Zealand's present-day rainforest flora. About 260 native species in 73 genera from 31 families are recognised and range from small ground-dwelling species and epiphytes to tall tree ferns.[9] Fossil evidence (mostly spores or impressions of fronds) suggests that ferns have a long and diverse history in the New Zealand region.[10]

Tiny fern spores are readily carried long distances by wind and water, and provide little information about where the source plants might have lived. As well, unrelated ferns may produce near-identical spores, making confident identification of isolated fern spores problematic. Most fronds, however, die while still attached firmly to the parent plant, and the chances of these being carried far are low.

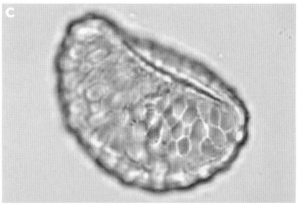

A|B The holotype or designated reference specimen of the fern *Davallia walkeri*, named in honour of Dr Alan Walker, who provided access to the site from 2000 to 2011. UK, JGC

C|D A light microscope photograph and an SEM image of spores from *Davallia walkeri*. JMB, EG

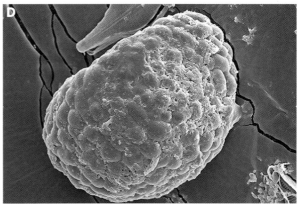

A|B A *Davallia*-like sterile frond and a specimen of the New Zealand endemic *Davallia tasmanii* for comparison. JGC

A

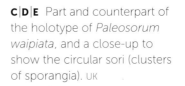

B

Not many organically preserved fronds have been found with in situ spores, although they are known from a few sites such as the late Eocene Pikopiko Fossil Forest and some South Island Miocene localities including Longford and Lake Manuherikia.[11] Therefore the discovery at Foulden of two fern fronds carrying spores has provided important new data on the history of pteridophytes in New Zealand. Identification and taxonomic placement are now possible.

To date, four fern macrofossils – two sterile and two fertile fronds – are known from Foulden. One fertile specimen is identified as a new species of *Davallia*. It has a three-pinnate frond (i.e. with three levels of branching) with free veins (the vein ends do not fuse together), no hairs on the lamina, stalked cup-shaped sporangia, elongate indusia (the spore-bearing areas and their flap-like covering) and spores characteristic of Davalliaceae.[12]

The family to which *Davallia* belongs is wide-ranging and largely epiphytic. Found mostly in Asia and Oceania, along with one species in Europe and two

C|D|E Part and counterpart of the holotype of *Paleosorum waipiata*, and a close-up to show the circular sori (clusters of sporangia). UK

C

D

E

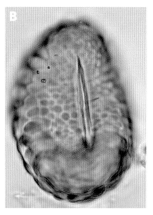

A Close-up of a sorus;
B|C close-up images of spores from the fossil under transmitted light and SEM. UK

BELOW Living *Lecanopteris* species from New Zealand for comparison with the fossil: **D|E** kōwaowao (*L. pustulata*); **F** mokimoki (*L. scandens*). JGC

in Africa and Madagascar, these ferns occupy warm temperate to tropical habitats/climates.[13] They live on branches or sometimes rocks, taking in water and nutrients from the host plant or rock surface and the air. This life habit, coupled with the ability of these fairly robust ferns to detach whole dead or dying fronds, makes them good candidates for preservation.

A single native species of *Davallia* occurs in northern New Zealand today. *Davallia tasmanii* is divided into two subspecies: one is known from the remote Manawatāwhi Three Kings Islands, where it is commonly found living on rocky substrates (a lithophyte); the other is known only from one site in Northland.[14]

The second fertile specimen is from an unrelated family, Polypodiaceae. It resembles species of *Lecanopteris* (formerly called *Microsorum*), a mainly tropical genus that has three species living in New Zealand today.[15] Like *Davallia*, *Lecanopteris* is an epiphyte and probably perched on rocky outcrops or branches overhanging the maar lake. However, because the fossil lacks some essential *Lecanopteris* characters, it was instead assigned to the fossil genus *Paleosorum*.[16]

CONIFERS

The presence of two extinct species of podocarp at Foulden Maar provides further evidence that New Zealand once had a much more diverse conifer flora than it has in modern times.

New Zealand is a centre of diversity of Southern Hemisphere conifers: there are 10 genera and 20 species in the modern flora, many of them ecological dominants in forests.[17] Some, such as kauri (Araucariaceae) and kahikatea, rimu, tōtara and matai (all Podocarpaceae) tower above the canopy. There has long been debate as to whether the conifers of New Zealand are ancient relicts or relatively recent arrivals.[18] The fossil record and recent research have now shown that not only are conifers ancient elements of the flora, but their diversity was once much higher.[19]

At Foulden, the pollen record lists 19 conifer taxa in the families Araucariaceae, Podocarpaceae, Taxodi-aceae and Ephedraceae, although none is identical to any living species. Seven have no close relatives in New Zealand today; they include relatives of the Huon pine and *Microcachrys*, which today live only in Tasmania. Pollen can travel long distances, and many of these conifers may have been living tens or even hundreds of kilometres away from the maar.[20]

The presence of two extinct species of Podocarpaceae in the diatomite is, however, unequivocal evidence that these trees formed part of the lakeside flora. The long strap-like leaves of *Podocarpus travisiae*, with their conspicuous single midvein and long petiole, are quite common. Often brownish in colour, they have distinctive cuticle with parallel rows of stomata. These leaves are much larger than any extant podocarp in New Zealand; their closest relatives are long-leaved *Podocarpus* species in the rainforests of Queensland (eastern Australia) and Southeast Asia.[21]

Less common are leaflets of a second podocarp, an extinct species of *Prumnopitys*. The leaves are narrow and pointed and the well-preserved cuticle is close to that of matai (*P. taxifolia*), a tall forest tree in the modern New Zealand flora. Similar fossils have also been reported from nearby Miocene localities.[22]

Podocarpus travisiae was the first plant fossil from Foulden Maar to be given a formal species name, by Mike Pole in 1993.
A A leaf of the 10cm-long fossil showing its long petiole. **B|C** Light microscope and SEM images of the cuticle of other specimens that display the rows of stomata typical of podocarps. These fossil leaves are more closely related to living large-leaved podocarps in Queensland (e.g. *P. grayae*) (**D**) than to the short, spiky leaves of *P. totara* (**E**), a local endemic tree species found around Foulden until fires destroyed the local vegetation. JMB, EG, JGC

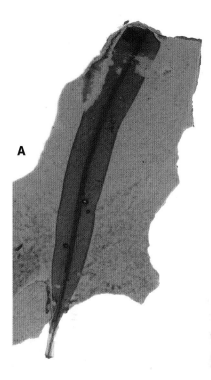

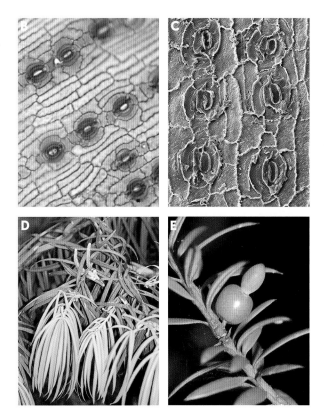

A|B|C Mataī-like *Prumnopitys* foliage from Foulden Maar, showing the short leaf shoot, single-veined leaf and characteristic cuticle for the genus. **D** A shoot of the living *P. taxifolia* is included for comparison. JGC, JMB

MONOCOTS

Although New Zealand today occupies a relatively small land area, it has one of the world's richest macrofossil records of monocots – flowering plants with a single embryonic leaf.

Monocots are a lineage representing around 22 percent of the flowering plants in which the seed contains a single embryonic leaf or cotyledon, distinguishing them from those with two cotyledons: 'basal' angiosperms (a diverse group of ancestrally divergent lineages with pollen that has a single aperture) and eudicots (a more derived evolutionary line representing around 75 percent of the flowering plants and with pollen that has at least three apertures). The largest monocot family is the orchids with around 25,000 species; but other familiar examples are grasses, palms, bananas, gingers, onions and asparagus. Many horticultural bulbs, such as lilies, tulips and daffodils, are also monocots.

At least 10 different types of monocot are known at Foulden from macro- and/or microfossils, and some are the first records in the world for their genus.[23] Compared to other flowering plants, the global monocot fossil record is sparse, largely because of their mostly herbaceous growth habit. Many taxa also tend to retain attached older leaves on the plant. Occasionally, fossil specimens are found with attached cuticle, and cuticle assessment can place the fossil into a family and genus. Unfortunately, because many monocot cuticles are fragile, this is relatively rare. Also, many monocots have generalised pollen that is produced by many different plant taxa.

Linear to lanceolate (narrow tapered) leaves with parallel veins are one of the most characteristic features of monocots – such as harakeke (New Zealand flax) and tī (cabbage trees). Those monocots whose older leaves drop off the plant are more likely to accumulate in fossil deposits.

Curiously, palms, the one group of monocots that usually fossilise well and that occur in warm rainforests, are so far completely absent as macrofossils at Foulden. However, their pollen is present, suggesting that fossilised leaves, flowers or fruits may be found in time.[24]

A few monocots run counter to the archetype of sword-like leaves with parallel veins. For example, *Ripogonum* and *Luzuriaga*, two genera that form part of the Foulden paleoflora, have broader, reticulate (net-veined) leaves that resemble those of typical eudicots.

ALSTROEMERIACEAE – *LUZURIAGA*

Luzuriaga (lantern berry, nohi, pūwatawata) is a low-growing, wiry semi-herbaceous perennial found throughout New Zealand. Its most characteristic feature, apart from its pretty white lantern-like flowers, is the unusual leaves with, literally, an odd twist. The stomata, which are generally on the lower surface of a

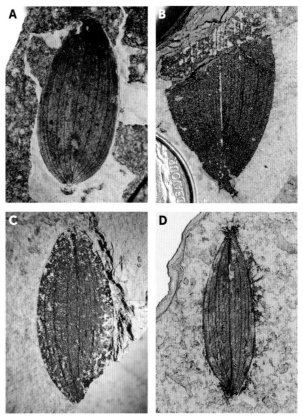

A|B|C|D Several leaves from Foulden of the extinct species *Luzuriaga peterbannisteri*. The species name honours the late Peter Bannister, former professor of botany at the University of Otago, who collected the type specimen. JGC, JMB

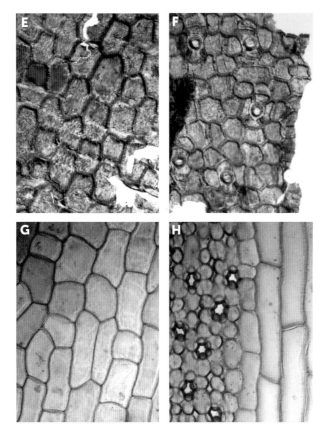

E|F Stained cuticles of fossil *Luzuriaga peterbannisteri* compared with two from the living New Zealand *Luzuriaga parviflora* (**G|H**). JMB

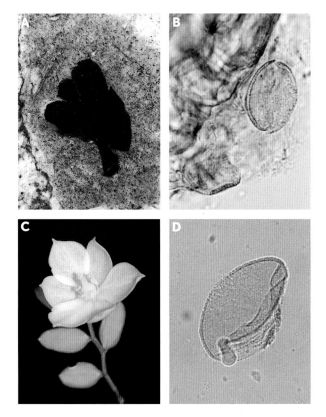

A|B Fossil flower and pollen grains of fossil *Luzuriaga peterbannisteri*, matched with a flower and pollen of *L. parviflora* (**C|D**). JMB, DCM, JGC

leaf, are here on the top; and the short petiole is then twisted so that the underside faces up to the light. Such leaves are termed 'resupinate'. The reason for this curious habit, found in only a few plants, remains a mystery to scientists.

Luzuriaga is one of four genera in the family Alstroemeriaceae, which includes the garden plant *Alstroemeria* (Peruvian lily) and the red-flowered climbing weedy herbaceous liana *Bomarea*, now widespread throughout New Zealand. Both were originally native to South and Central America. *Luzuriaga* also has South American connections, with three species in Chile. The remaining closely related genus, *Drymophila*, is found only in south-eastern Australia. These distributions hint at complex ancient dispersal patterns across some of the southern continents in the distant past.

The only pre-Quaternary fossil record for *Luzuriaga* – and the only macrofossil record in the world for the genus and the family – comes from Foulden.[25] The small, distinctive ovate leaves show two orders of veins and a conspicuously resupinate petiole. Flowers of *Luzuriaga* have also been found. Although dark and not very flower-like at first glance, they proved to contain fragile pollen grains of *Liliacidites contortus*, a close match to the pollen of the endemic local species.

Luzuriaga parviflora grows either as an epiphyte or as a low-growing semi-herbaceous perennial herb in deep litter on the forest floor, in moss beds or on the margins of swamps. For leaves and fragile flowers to have fallen into the lake, the *Luzuriaga* at Foulden must have grown close to the edge, perhaps in moss beds. These fossils also highlight ancient biogeographic connections between New Zealand, Chile and Tasmania.

ASTELIACEAE – *ASTELIA*

The characteristics of some narrow leaf fragments found at Foulden suggest they might belong to the herbaceous genus *Astelia* (also known as bush lilies, perching lilies, kōkaha or kahakaha), which is common throughout New Zealand.[26]

A *Astelia fragrans* growing as an epiphyte on a rainforest tree trunk, showing typical short stems bearing clumps of long, narrow, sword-like leaves and a terminal inflorescence with brightly coloured berry-like capsules.

B One of the more conspicuous features in *Astelia* is the abundant fine hairs on the leaf surfaces, which gives them a silky, often white or silvery appearance. These hairs have conspicuous multicellular bases (**C**). JGC

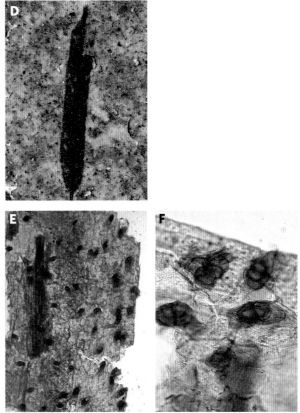

There are about 25 *Astelia* species extending from New Zealand and Australia through the Pacific to Hawai`i, Argentina, Chile and the Falkland Islands. Today, there are two main ecological groups within *Astelia*: warm temperate to subtropical, and cool temperate to alpine.[27] Some are epiphytes, growing as clumps high on tree branches after a seed has germinated in a cavity. Most are insect-pollinated (although the pollen of some epiphytes has also been found in bat droppings), and they typically have bright red or orange fleshy, berry-like fruits that are dispersed by birds.[28] The *Astelia* plants at Foulden were likely to have been epiphytes, perching above the lake on branches from which they may have been dislodged during storms.

D|E|F Part of a narrow, elongate leaf showing the prominent midrib extending beyond the leaf lamina. These fragmentary fossil leaves differ from all living species of *Astelia* and were given the species name *antiqua*, in reference to their age. Staining with crystal violet enhances the cuticular features. JMB

A|B|C *Cordyline* has large sturdy leaves that drop off as they age, making it an excellent candidate for preservation in the fossil record. The discovery of large fragments of leaves at Foulden, possibly representing several species, is proof of this. JGC

ASPARAGACEAE – CABBAGE TREES

The iconic cabbage tree (*Cordyline* spp.) occupies forest edges and riparian areas of open rainforest across New Zealand, though generally prefers more fertile soils. It is also widespread in eastern Australia and its distribution extends to Chile, Papua New Guinea, New Caledonia and as far as the Mascarenes Islands in the Indian Ocean. No *Cordyline* macrofossils have yet been described formally from New Zealand, but fossils of the related extinct genus *Paracordyline* are known from the Eocene of Australia and the Oligocene of the remote subantarctic Kerguelen Islands in the southern Indian Ocean.[29]

The sculpturing on the cuticle of the larger-leaved Foulden *Cordyline* fossil (**D**) is similar to that of an Oligocene-aged fossil from the Kerguelen Islands (**E**) and also the extant New Zealand *C. indivisa* (broad-leaved cabbage tree) (**F**); but the Foulden fossil differs in enough features for it to be regarded as a new species. JGC, JMB

ORCHIDACEAE – ORCHIDS

Orchids are cosmopolitan and diverse and many species possess showy exotic flowers, but they are exceptionally rare as fossils. Despite there being some 25,000 living species, there are only 10 fossils, two of which are from Foulden. Of these 10, only one possible fossil orchid flower is reported, from the Northern Hemisphere.[30] In 2007 a pollinium (pollen mass) from an orchid flower was discovered attached to an extinct stingless bee imprisoned in Dominican amber, from the Dominican Republic.[31] Since then, further pollinia have been reported, attached to a fungus gnat preserved in Baltic amber, and on beetles and/or bees from Mexican and Dominican amber.[32]

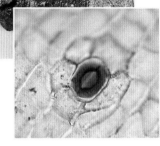

Fossil leaves and cuticles of *Dendrobium winikaphyllum* (**A**) and *Earina fouldenensis* (**B**). JGC, JMB

The leaves proved to be closely related to two different genera still found in New Zealand: *Dendrobium* (**C**) and *Earina* (**D**). JGC

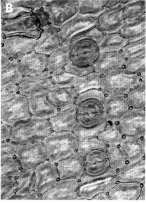

The leaves of both *Earina* (**A**) and *Dendrobium* (**B**) are tough, shed in abundance, and their cuticles show patterns characteristic for the genera.
JGC, JMB

Although disputed orchid fossils have been reported as far back as 1857,[33] there were no definitive vegetative orchid fossils until 2008, when the research team collected some lanceolate leaves with parallel venation at Foulden. Drab in colour, they looked less than impressive, but careful preparation showed distinctive features of the cuticle that could only belong to an orchid. To confirm this, we sent photographs of leaves and cuticles of the fossils and living orchid species to monocot experts at the Royal Botanic Gardens, Kew, who agreed that these were indeed the first and second global records of orchid leaves. The fossils were named *Earina fouldenensis* and *Dendrobium winikaphyllum* in 2009.[34]

As the first fossils from the largely epiphytic Southern Hemisphere subfamily Epidendroideae, these provided important calibration points for the molecular phylogeny (the DNA-based family tree) of Orchidaceae. Until about 2000 it was not known whether orchids were fairly recent arrivals in the plant world, or if they had first appeared far back in the Cretaceous but just lacked a fossil record. The Foulden fossils helped to confirm the latter: orchids radiated and dispersed globally long before 23 Ma, probably as far back as the early late Cretaceous (c. 105–76 Ma).[35]

Orchids have tiny wind-blown seeds and possess an unusual arrangement of pollen grains, whereby several tiny grains cluster together to form a pollinium that is then transferred as a single unit during pollination.

Epiphytic orchids are most diverse in tropical regions; and terrestrial species – those that grow on the ground – are generally more common in temperate areas. Modern New Zealand has more than 160 orchid species in 26 genera, but only eight species are epiphytic, and none possess the large showy flowers of their tropical cousins.

The winikā or Christmas orchid (*Dendrobium cunninghamii*), the most impressively flowered New

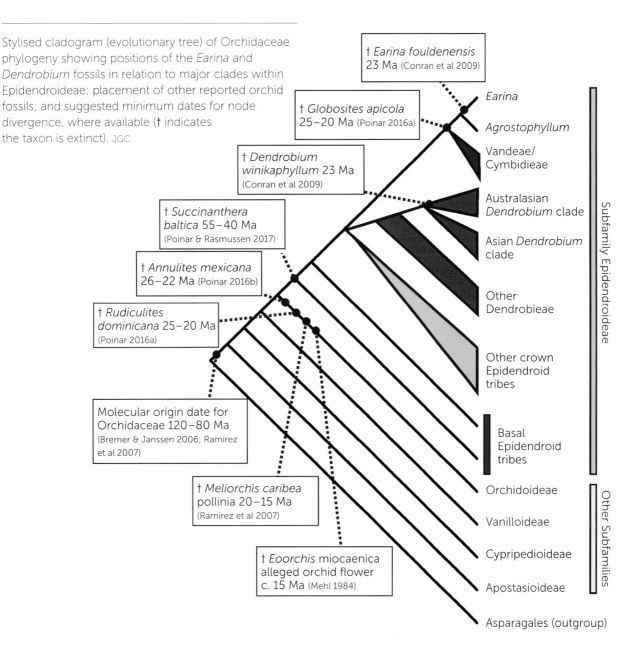

Stylised cladogram (evolutionary tree) of Orchidaceae phylogeny showing positions of the *Earina* and *Dendrobium* fossils in relation to major clades within Epidendroideae; placement of other reported orchid fossils; and suggested minimum dates for node divergence, where available († indicates the taxon is extinct). JGC

† *Earina fouldenensis* 23 Ma (Conran et al 2009)

† *Globosites apicola* 25–20 Ma (Poinar 2016a)

† *Dendrobium winikaphyllum* 23 Ma (Conran et al 2009)

† *Succinanthera baltica* 55–40 Ma (Poinar & Rasmussen 2017)

† *Annulites mexicana* 26–22 Ma (Poinar 2016b)

† *Rudiculites dominicana* 25–20 Ma (Poinar 2016a)

Molecular origin date for Orchidaceae 120–80 Ma (Bremer & Janssen 2006; Ramírez et al 2007)

† *Meliorchis caribea* pollinia 20–15 Ma (Ramírez et al 2007)

† *Eoorchis* miocaenica alleged orchid flower c. 15 Ma (Mehl 1984)

Earina

Agrostophyllum

Vandeae/ Cymbidieae

Australasian *Dendrobium* clade

Asian *Dendrobium* clade

Other Dendrobieae

Other crown Epidendroid tribes

Basal Epidendroid tribes

Orchidoideae

Vanilloideae

Cypripedioideae

Apostasioideae

Asparagales (outgroup)

Subfamily Epidendroideae

Other Subfamilies

Zealand epiphytic orchid, as well as two of the three native *Earina* species, are relatively common epiphytes or lithophytes in New Zealand forests and occupy a wide range of moist habitats.[36] Their leaves are tough and shed in abundance, making them ideal candidates for fossilisation. The Foulden species probably lived on overhanging branches or rocky outcrops around the edge of the lake.

The discovery of orchid leaves at Foulden was a major breakthrough. It showed that epiphytic orchids had expanded into Zealandia by the mid-Cenozoic, and it demonstrated the importance of southern continents in the diversification of Orchidaceae.

To date, no orchid flowers or pollen have been found at Foulden.

RIPOGONACEAE – SUPPLEJACK

Leaves of another monocot, supplejack (*Ripogonum*, kareao), are quite common at Foulden.[37] These are sometimes large (up to 18cm long) and have distinctive reticulate veins between three levels of nerves. There appear to be at least two species present in the maar.

Supplejack in modern New Zealand forests has tough, woody, loosely coiled entangling stems that can grow more than 10m to reach sunlight. With long-lived, thick and tough leaves and stems, the lianes sometimes form almost impenetrable barriers. The curious name is used for similarly flexible though unrelated plants in other parts of the world. Māori and early European settlers made extensive use of kareao stems in woven items such as fish traps and baskets, and the plant was noted for its medicinal properties. The young asparagus-like shoots are also edible.[38]

The genus *Ripogonum* was created in 1776 by Johann and Georg Forster, the father and son naturalists aboard the *Resolution* on Captain James Cook's second voyage to New Zealand.[39] A single red-fruited species, *Ripogonum scandens*, is common in lowland podocarp and broad-leaved forest in the North and South islands, Rakiura Stewart Island and Rēkohu Chatham Islands, although the Chatham Island form is genetically different and may represent a separate taxon.[40] The berries are eaten and seeds readily dispersed by the kererū (New Zealand pigeon), kōkako and tūī.

Species of *Ripogonum* are all typically vines or lianes of wet forests and often grow near watercourses. Other species, including some that closely resemble the New Zealand fossils, occur in Australia and Papua

A|B Two fossil species of supplejack (*Ripogonum*) from Foulden Maar. The thick tough leaves make them good candidates for fossilisation. This *Ripogonum*-like 13cm flattened branchlet (**C**) was still flexible and supple despite being 23 million years old! JGC

D *Ripogonum scandens*, showing the wiry climbing stems, opposite leaf arrangement and panicles of small, green starry flowers. **E** The leaf X-ray shows the strong main vein, divergent secondary veins and weak tertiary veins close to the leaf margin.
F The stained cuticle shows epidermal cells with strongly sinuous cell walls resembling interlocking jigsaw-puzzle pieces, and stomata. **G** Wiry stems of the vine, which resemble the fossil stem piece recovered from Foulden. JGC, JMB

E

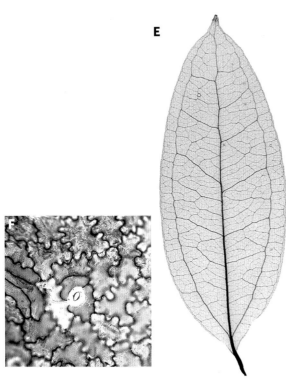

F

G

New Guinea, and extinct fossil species are also known from the Eocene of New Zealand, South America and Tasmania.[41]

In a geological timeframe, *Ripogonum* has a very long fossil record in New Zealand: recently discovered fossils from Otaio Gorge in South Canterbury extend the record back to 50 Ma in the early Eocene. Similar *Ripogonum* fossils are found in strata of early Eocene age in Tasmania and South America, making the genus of considerable Southern Hemisphere biogeographic interest.[42]

LAKE-EDGE VEGETATION

The maar must have had some areas in which wetland plants flourished. Pollen from bur reeds (*Sparganium*), raupō or bulrushes (*Typha*), harakeke (*Phormium*), oioi or jointed rushes (Restionaceae), sedges (Cyperaceae) and the aquatic eudicot herb water milfoil (*Myriophyllum*) suggest there were some shallow swampy margins to the generally steep-sided crater lake walls.

Typhaceae fossils are known from New Zealand mainly as pollen related to the living genera *Typha* and *Sparganium*.[43] Several raupō leaf macrofossils have now been found at Foulden, and leaves and occasional seeds have been reported previously from several Miocene New Zealand localities.[44]

The modern New Zealand species are perennial aquatic plants that form extensive colonies in marshy areas. Raupō roots are permanently submerged, and their creeping, starchy underground stems were a food source for Māori. The long narrow leaves are green in spring, dying to orange in autumn.[45]

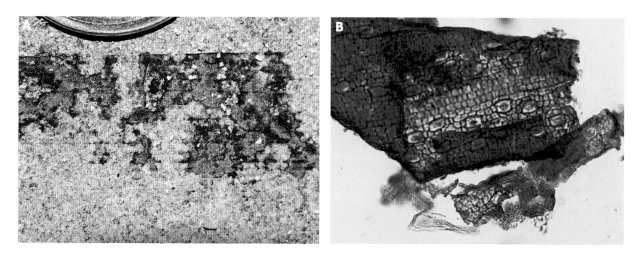

Leaf fragment (**A**) and epidermal cuticle (**B**) of *Typha* (raupō) recovered from Foulden Maar. JGC, JMB

C|D Modern raupō
(*Typha orientalis*) growing
on the edge of a swamp
in Central Otago.
E A leaf epidermal
preparation showing
stomata similar to those
of the fossil (**A**|**B**). DEL, JMB

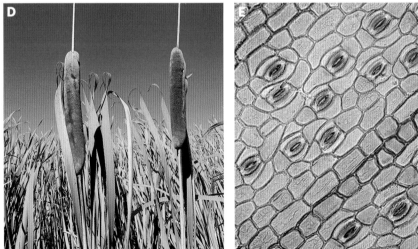

CHAPTER 5

FOSSIL PLANTS 2: FLOWERING TREES, SHRUBS AND FUNGI

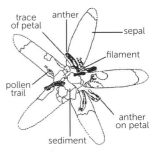

Few fossil sites have several elements of the same plant preserved, but maar lakes are the exception. The fossil treasures discovered at Foulden Maar include many flowering plants that have associated parts preserved. Bark and wood are uncommon, but leaves with cuticle, flowers with pollen, fruits and seeds from the same plant species are preserved. Such associations are extraordinarily rare in the fossil record globally.

In addition to the monocot fossils described in Chapter 4, there are also numerous basal angiosperm and eudicot fossils from the site. Many of these represent the first fossil record for their genus or family. Taken together, these fossils showcase the wide diversity and biogeographic connections of the New Zealand flora at the Oligocene–Miocene boundary 23 million years ago, with links to Australia, New Caledonia and South America all represented at the paleolake.

Many new species and at least one new genus of extinct flowering trees and shrubs have come to light at Foulden Maar, including *Fouldenia staminosa*, New Zealand's first fossil flower with petals, anthers and pollen preserved.

FLOWERING TREES AND SHRUBS

FOULDENIA STAMINOSA

Fossil flowers, because of their fragility, used to be rare discoveries; examples with pollen preserved in their anthers are still rare. A paper published in 1995 listed just 40 examples of pollen-bearing fossil flowers of Cenozoic age worldwide, and only six of Miocene age.[1] Hundreds more have been reported since then, many of them extracted as three-dimensional fossils from charcoalified sediments using new techniques. The record now extends far back to the early Cretaceous and the origins of flowering plants.[2]

Until 2003, the list of fossil flowers from New Zealand was restricted to one specimen from the Miocene Lake Manuherikia deposit and late Cretaceous flower-like reproductive structures.[3] So the discovery at Foulden Maar of an isolated, fragile, but readily recognisable compressed five-petalled flower with pollen preserved in its anthers made a stir in the paleobotanical world.

But this first discovery, although beautifully preserved, was challenging to identify.[4] To describe a fossil flower we need to identify all characteristics of the

OPPOSITE The discovery of *Fouldenia staminosa* at Foulden Maar made news among paleobotanists worldwide. The flower is 23mm across, with one whorl of five sepals joined only at the base. It is radially symmetrical (star-shaped). The sepals are thick, oval and 10mm long, and the corolla consists of five free petals that are much shorter than the sepals. It had either five or 10 large prominent stamens and thick-walled anthers. JMB

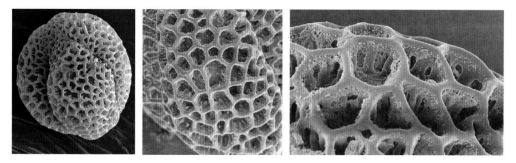

SEM images of *Rubipollis oblatus* pollen extracted from the *Fouldenia staminosa* anthers. This pollen type had already been described from dispersed pollen grains at Oligocene and Miocene sites around New Zealand, but had not been linked previously to a flower. JMB, EG

A|B A partial inflorescence of *Akania gibsonorum*, showing numerous 5mm-wide flowers. For comparison, **C** is a flower of the modern species. The tiny unspecialised fossil flowers suggest that *A. gibsonorum* was probably pollinated by small insects. UK, JGC

flower and its associated pollen and then, if possible, assign it to a family and a genus. In this case, however, no exact match could be found in the modern New Zealand flora or in those of nearby New Caledonia or Australia, so we concluded that this flower represented a new extinct genus and species. It was named *Fouldenia staminosa* – the generic name to honour Foulden Maar, and *staminosa* for its distinctive stamens. There are no associated leaves, and although 17 further years of searching have yielded another 40 plus fossil flowers, there are no other examples of this one.

It is possible that the flower belongs to a tree or shrub in the family Rutaceae, which includes lemons. We cannot be sure of its colour in life, but the flower architecture suggests that it was probably insect-pollinated. Tubular or bell-shaped flowers are designed for birds, such as honeyeaters, to access, whereas small insignificant flowers that produce copious amounts of pollen are more likely to be pollinated by wind and/or small pollen-feeding insects.[5]

AKANIACEAE: AN AUSTRALIAN RELICT IN THE NEW ZEALAND MIOCENE

Another intriguing fossil flower, this one found at Foulden in 2017, has an unusual distribution. It is a paniculate inflorescence – a branched cluster of flowers of which the sepals, petals and anthers are fused onto a cup-like structure with a three-chambered ovary sitting inside. Study of the flower and its pollen confirmed that this was the first fossil flower of *Akania* (Akaniaceae), found today only in part of coastal eastern Australia. This new extinct species has been named *Akania gibsonorum*.[6] The floral structure and pollen resemble those of the only modern species, *A. bidwillii*.

Sometimes called turnipwood, Akaniaceae are a family of flowering rainforest trees with just two genera, each with a single species: *Akania*, endemic to Australia; and *Bretschneidera*, known from China, Vietnam and India. *Akania bidwillii* is an uncommon small tree up to 12m tall with large, toothed compound leaves. It grows in subtropical/warm-temperate rainforests on the coastal lowland and adjacent ranges of eastern Australia, from northeast New South Wales to the Sunshine Coast in southeast Queensland.

No fossils of *Akania* have been reported from Australia, but fossil leaves with its distinctive marginal tooth morphology and venation, as well as Akaniaceae wood, have been described from South America, suggesting an origin there with subsequent dispersal to New Zealand and Australia and on into Asia.[7]

None of the thousands of leaves collected from Foulden so far have the distinctive venation typical of *Akania* leaves, although we will keep looking for them. The unexpected discovery of the inflorescence at Foulden suggests that *Akania* was once widespread in South America and New Zealand as well as in Australia, and its present restricted distribution reflects shrinkage of a much broader geographic range.

ALSEUOSMIACEAE – *ALSEUOSMIA*

Two flowers from Foulden are a close match for those of the New Zealand endemic genus *Alseuosmia* (toropapa; Alseuosmiaceae). They have a long funnel-like corolla and five stamens inserted near the throat.

There are eight or so species of this small, evergreen understorey shrub of broad-leaved forests in the North Island and northern South Island today.[8] The flowers, which are pollinated by birds or moths, are bell-shaped, often brightly coloured, with conspicuous stamens. No fossil leaves or cuticle that might belong to this genus have yet been identified; however, the leaves of *Alseuosmia* are hugely variable in size and shape, mimicking those of many other plants, and the cuticle has no clearly distinguishing features. These flowers are the only known fossils of *Alseuosmia*.

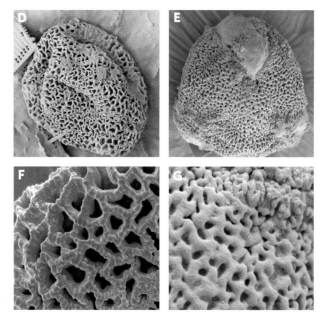

The large tricolpate (having three apertures) pollen grains shown in SEM images for the fossil (**D**|**F**) and the living *Akania* species (**E**|**G**). JMB, EG

A|**B** Two flowers with conspicuous stamens at the mouth of the floral tube that are a close match to *Alseuosmia*. Flowers of *A. quercifolia* (**C**) and *A. macrophylla* (**D**) for comparison. JGC, BLL

Fossil Araliaceae from Foulden Maar: five-finger leaf and fruits (**A**|**B**) and *Meryta* (**C**). TRR, JGC, JMB

D|**E**|**F** New Zealand *Pseudopanax arboreus* (puahou) leaves and fruit and a *Meryta sinclairii* (puka) leaf for comparison. JGC

ARALIACEAE − *PSEUDOPANAX* AND *MERYTA*

Some leaves at Foulden are larger than most seen in the modern New Zealand flora, at least on the mainland. Leaves and berry-like fruits of the widespread New Zealand five-finger and horoeka, lancewood (*Pseudopanax*) are relatively common as fossils at Foulden Maar, suggesting that birds dispersed the seeds at the site.[9] One particularly big fossil leaf is a close match to *Meryta*, a native tree naturally restricted to Manawatāwhi Three Kings Islands but now planted widely as a handsome ornamental in warmer areas of New Zealand. Only one species, puka (*M. sinclairii*), exists in New Zealand, but other species are found in Queensland, New Caledonia and other Pacific islands.[10]

ATHEROSPERMATACEAE – *LAURELIA*

Laurelia (Atherospermataceae) is a member of a small family with seven genera and 14 species restricted today to temperate rainforests of Australia, New Caledonia, the mountains of Papua New Guinea, Chile and New Zealand. The local species, pukatea (*L. novae-zelandiae*), is a tall tree (up to 35m) with leaves that have variable marginal serration. In warmer northern regions and as far south as Nelson, pukatea is often found growing in poorly drained sites. It is one of very few New Zealand trees that possess 'plank buttresses' (thin triangular flanges extending up the trunk and along the roots) and pneumatophores (above-ground roots), both features more typical of trees in tropical forests.[11] The single-seeded fruits or achenes are wind- or possibly bird-dispersed by way of distinctive hairy tufts.

Leaves found at Foulden that are variable in size, type of margins and venation angles have been attributed to a new fossil species, *Laurelia otagoensis*.[12] *Laurelia* achenes from Foulden are the first fossil fruits recorded for the family.

A|B The holotype of *Laurelia otagoensis* and a hairy fruit (achene). **C|D** An extant New Zealand pukatea (*L. novae-zelandiae*) leaf and achene for comparison. JMB, JGC

Leaf epidermal cuticles stained blue with crystal violet. **E|F** Lower surface of fossil displaying stomata and a close-up of a thickened hair base. **G|H** Lower surface of *L. novae-zelandiae* showing same for comparison. JMB

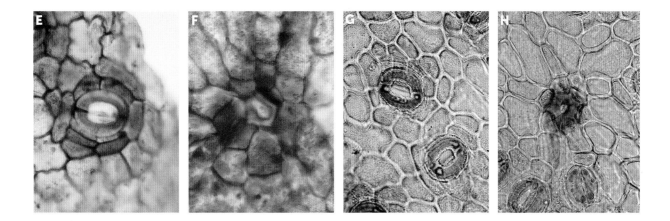

BIGNONIACEAE

Several examples of a distinctive subrounded, flat two-winged seed have been recovered from Foulden. They are likely to belong to an extinct liane or tree within Bignoniaceae, a widespread pantropical family of trees and vines.[13] The single living New Zealand member of the family, the Three Kings vine (*Tecomanthe speciosa*), is endemic to Manawatāwhi Three Kings Islands and has the dubious distinction of having been reduced by browsing goats to a single plant when it was discovered in 1945. Fortunately, it is self-fertile, grows readily from cuttings, and is now cultivated widely in the warmer parts of New Zealand as a rather vigorous ornamental climber.[14]

The fossil seed, although similar in overall form and embryo shape, is a bit different from this and possibly represents either another now extinct *Tecomanthe*-like species or perhaps a relative, the tropical tree *Deplanchea* (golden pagoda tree), which has species in Australia and New Caledonia.[15]

A Fossil two-winged Bignoniaceae-like seed from Foulden; **B|C** Foliage and flowers of the Three Kings vine *Tecomanthe speciosa*; **D** Seeds of the Australasian tree *Deplanchea speciosa* and Australian vine *Tecomanthe burungu* for comparison, both showing two wings, a central rounded area and heart-shaped embryo. Seeds of the living New Zealand species are less like the fossil. JGC, Gildas Gâteblé and Frank Zich

CELASTRACEAE – 'OLD GOLD'

Some commonly collected leaves at Foulden are known as 'Old Gold' for their characteristic yellowish-brown colour, distinctive venation and wrinkled surface. They have well-preserved distinctive cuticle, with a box-like arrangement of subsidiary cells surrounding the stomata. They have been assigned to Celastraceae (bittersweet), a worldwide family no longer present in New Zealand but with several genera in Australia and New Caledonia.

A|B A fossil 'Old Gold' leaf and cuticle. **C|D** A leaf and cuticle from the modern Australasian genus *Denhamia* for comparison. JGC, JMB

CUNONIACEAE – *ACKAMA* AND *WEINMANNIA*

Cunoniaceae are a diverse and ancient Southern Hemisphere rainforest tree and shrub family from New Zealand, Australia, Papua New Guinea, New Caledonia, the Pacific, South America and Madagascar, with 27 genera and over 300 species.[16] Four endemic species from two genera, *Ackama* and *Weinmannia*, are present in the modern New Zealand flora (two species each). One of these – kāmahi or *Weinmannia racemosa* – is thought to be the most common native tree.[17]

The small flowers of Cunoniaceae are pollinated by insects and birds and the seeds are wind-dispersed.[18] The family has a fossil record extending back to the Late Cretaceous, and fossil Cunoniaceae leaves, detached leaflets and reproductive structures have been found at Foulden, many with excellent cuticle.[19] Flowers related to *Ackama* and/or *Weinmannia* and capsular *Weinmannia*-like fruits are also preserved.

A|B A fossil leaflet from a compound *Weinmannia* leaf and a two valved capsule. **C|D** A compound leaf and capsules of tōwai (*W. silvicola*) for comparison. JMB, JGC

E|F Fossil Cunoniaceae flower with in situ pollen. The flower has four or five perianth segments that are united at the base, 8–10 stamens and a two-chambered ovary. Pollen from the flower is tricolporate. Flowers of makamaka (*Ackama rosifolia*) (**G**) and tōwai (*Weinmannia silvicola*) (**H**) for comparison. JMB, JGC

ELAEOCARPACEAE – *ELAEOCARPUS* AND/OR *SLOANEA*

The family Elaeocarpaceae, with over 600 species in 12 genera, are rainforest trees and shrubs that occur throughout the warmer regions of Southeast Asia, Australia, New Caledonia, Madagascar and Central and South America, with some genera extending to temperate Chile and New Zealand.[20] The largest and most widespread genera are *Elaeocarpus* (c. 350 spp.) and *Sloanea* (c. 150 spp.), and the family has an extensive fossil record largely based on the fossilised fruits (endocarps) of *Elaeocarpus*.[21]

Flowers and leaves with affinities to *Elaeocarpus* or *Sloanea* have been reported from Foulden, including one leaf with fossilised scale insects attached.[22] Two species of *Elaeocarpus* – hīnau (*E. dentatus*) and pōkākā (*E. hookerianus*) – are small trees found throughout New Zealand. *Sloanea* is a closely related genus found mainly in Australia and South America but absent from the New Zealand flora today. These flowers are pollinated by small insects and the fruits are dispersed by birds.

One pollen-bearing Elaeocarpaceae flower from Foulden is not related to any living New Zealand species but resembles *Dubouzetia* from New Caledonia, Papua New Guinea and northern Australia, and shares distinctive pollen features with the temperate South American genus *Crinodendron* (lantern tree). However, the pollen of *Dubouzetia* needs detailed examination and the precise affinities of this fossil are the subject of ongoing research.

A|B Bell-like Elaeocarpaceae flowers displaying features similar to *Elaeocarpus* and/or *Sloanea*. Flowers of pōkākā (*E. hookerianus*) (**C**) and the Australian blueberry ash (*E. reticulatus*) (**D**) for comparison. JMB, JGC

E|F The second Foulden elaeocarp flower is radially symmetrical, with many stamens apparently surrounding a disc, and is either a male flower or the female parts are not preserved. The stamens have long anthers and short filaments. Structurally, the flower resembles the Australian genus *Dubouzetia* and the pollen is well preserved with very distinctive ornamentation similar to *Crinodendron*. Flowers of *Dubouzetia guillauminii* (**G**) from New Caledonia and *Crinodendron hookerianum* (lantern tree) (**H**) for comparison. JMB, EG, Gildas Gâteblé, JGC

EUPHORBIACEAE – *MALLOTUS* OR *MACARANGA* = *MALLORANGA*

Some broadly ovate to round leaves with distinctive venation that are quite common at Foulden are unlike any plants found in New Zealand today. They are most similar to species in the *Mallotus–Macaranga* complex of the spurge and rubber tree family (Euphorbiaceae), which are common in moist secondary forests in Southeast Asia. In the subtropics of eastern Australia they often also occupy seasonally dry rainforest margins north of 32°S.

The features that enabled identification of these fossil leaves are subtle, and since we could not be certain whether *Mallotus* or *Macaranga* were present in New Zealand in the early Miocene, we proposed a new generic name combining both: *Malloranga*.[23]

The leaves have unusual disc-shaped glandular scales and nectar-secreting plant glands outside the flowers to attract insects, such as ants, which then protect the leaves from attack by herbivores. They also have tiny hair tufts in the main vein of the lower leaf surface, which act as homes for even tinier mites that help protect the leaves from predators and remove fungal spores and detritus.

In addition to the numerous *Malloranga* leaves at Foulden Maar, we also found an inflorescence with male flowers bearing *Nyssapollenites endobalteus* pollen; and three-lobed capsular fruits assigned to *Euphorbiotheca mallotoides*. All of this supported the association with *Mallotus/Macaranga*, and more recent study has suggested an affinity with *Mallotus nesophilus* from northern Australia.[24] This suggests wind pollination and bird dispersal of the seeds, although *Mallotus* has recently been shown to be pollinated by both wind and insects.[25]

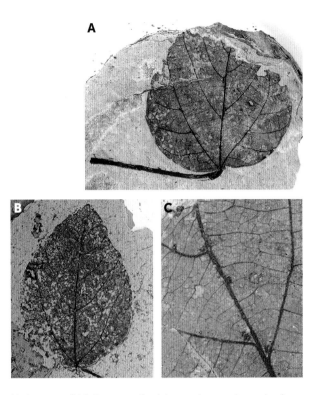

Holotype of *Malloranga fouldenensis*: **A** a large leaf with three main veins and a long petiole; **B** a complete leaf; **C** a closeup of hair-tuft domatia in the leaf vein axils. The cuticle is not well preserved, suggesting that the leaves were thin. JMB

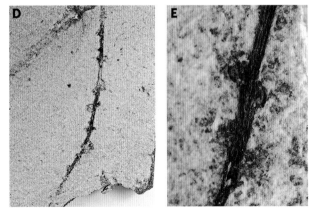

D|E|F|G In addition to the leaves, there was also a tiny inflorescence of male flowers bearing the dispersed pollen type *Nyssapollenites endobalteus*. JMB

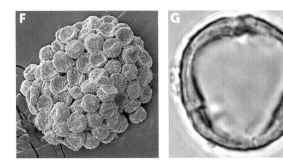

LAURACEAE

Thousands of leaves have been collected from the small exposures of diatomite at Foulden Maar, and cuticles of more than 800 have been prepared so far. These have revealed that the rainforest immediately around the small lake was dominated by many genera and species of Lauraceae, a large family of usually evergreen trees and shrubs growing mostly in Australia, Southeast Asia and tropical America. The modern New Zealand flora has five species in three genera, mostly in northern New Zealand, but the predominance of lauraceous leaves at Foulden shows that they were once also common and more diverse in the south.[26]

Without the evidence from leaf fossils we could not know this, for no traces have been found of the thin-walled fragile pollen of Lauraceae.[27] Thus, a study based only on the pollen from Foulden would present a biased and completely erroneous picture of the early Miocene vegetation at the site.[28]

Examples of Lauraceae leaves from Foulden Maar.
A *Cryptocarya*-like *Laurophyllum longfordensis*;
B *L. microphyllum*; **C** *L. taieriensis*; **D** *Beilschmiedia*-like *L. lacustris*; **E** *L. vulcanicola*; **F** *Litsea*-like *Laurophyllum calicarioides*. JMB, JGC

BIRDS AND OTHER VERTEBRATES

While no vertebrate fossils apart from fish have been found at Foulden so far, numerous remarkable discoveries have been made in the past two decades at fossil sites of early to mid-Miocene age at nearby St Bathans. The isolated bones of a wide range of birds have been collected, including those of a large extinct fruit pigeon.[29] Others are mostly from waterfowl (ducks and geese), but parrots, pigeons, gulls and many other species have been found along with moa eggshell and the bones of bats, skinks and turtles. The most charismatic fossils discovered thus far are the remains of a small freshwater crocodile and the oldest New Zealand tuatara fossil. These dispersed and fragmentary bones were deposited in rarely preserved deltaic environments on the shores of the huge inland Lake Manuherikia.[30]

Modern New Zealand native Lauraceae are represented by two leafless vines in the genus *Cassytha*, as well as mangeao (*Litsea calicaris*), tawa (*Beilschmiedia tawa*) and taraire (*B. tarairi*), three rainforest canopy trees that grow mainly in warmer areas of the North Island at low to medium altitudes (tawa extends as far south as Kaikōura). Lauraceae are also common in the rainforests of Australia (115 species in eight genera) and New Caledonia (47 species in five genera), where they are often canopy dominants.

The excellent preservation of leaves has made it possible to distinguish at least 10 species of Lauraceae at Foulden.[31] Placed into the genus *Laurophyllum*, they represent four leaf taxa with affinities to *Beilschmiedia*,

G|H Taraire (*B. tarairi*) flower and fruiting branch. JGC

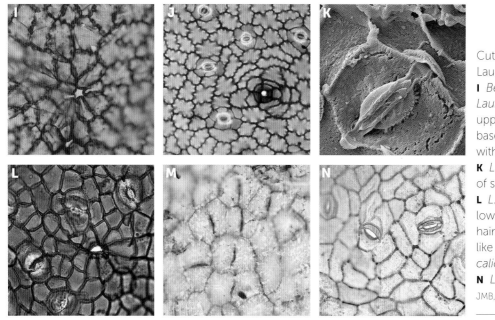

Cuticle examples of Lauraceae leaf fossils: **I** *Beilschmiedia*-like *Laurophyllum sylvestris* upper cuticle with hair base; **J** lower cuticle with stomata; **K** *L. lacustris* close-up of stomata under SEM. **L** *L. microphyllum* lower cuticle with hair base; **M** *Litsea*-like *Laurophyllum calicarioides* upper; **N** *L. calicarioides* lower. JMB, EG

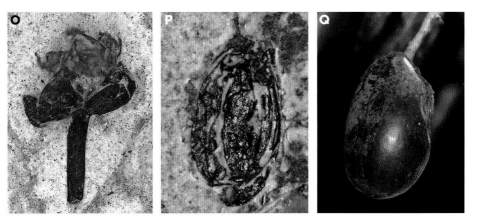

O The bracteate fossil flower head closely resembles those of the extant *Litsea calicaris*, but as some floral details are indistinct in the fossil, it is placed into the form genus *Litseopsis*. **P** Lauraceae fruit from Foulden and a *Beilschmiedia tawa* drupe (**Q**) for comparison. JMB, JGC

one that is close to *Litsea calicaris*, and five that are most closely related to the widespread tropical genus *Cryptocarya*, no longer found in New Zealand.

The compressed ovoid fruits are good matches for the fleshy fruit of *Beilschmiedia* and *Cryptocarya*. Although no bird fossils have yet been found at Foulden, the relative abundance of fossilised large-seeded Lauraceae fruits at the site suggests that sizeable fruit-eating dispersers have a long history in New Zealand.[32] Today, *Beilschmiedia* seeds are dispersed by kererū and kōkako.

LORANTHACEAE – MISTLETOE

A distinctive, single, detached fossil bell flower found at Foulden was determined likely to be an extinct relative of the New Zealand endemic yellow mistletoe or pirita (*Alepis flavida*). This identification is further supported by the presence of the distinctive pollen type, *Gothanipollis perplexus*, a fossil mistletoe previously known only from dispersed pollen.[33]

Mistletoes are aerial parasites, meaning they take their nutrition from the host plant on which they perch. There are five native genera (four endemic) and six species of mistletoe still present in New Zealand, plus one recently extinct genus. Several of these are widespread and have colourful flowers that bear 'explosive' pollen. They are pollinated variously by birds, bees and wasps, and are dispersed primarily by birds attracted to their abundant, sticky-seeded and fleshy berries.[34]

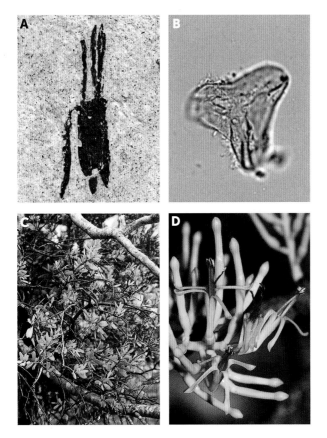

A|B A fossil mistletoe flower and pollen grain of *Gothanipollis perplexus*. For comparison, yellow mistletoe (*Alepis flavida*) (**C**) on a *Nothofagus* host tree (green *Alepis* foliage in the foreground) growing together with scarlet mistletoe (*Peraxilla tetrapetala*), and (**D**) a close-up of *Alepis* flowers. JMB, JGC

MELIACEAE – 'DYSOXYLUM'

The present-day New Zealand flora includes many evergreen tree and shrub genera and species that are mostly restricted to warmer northern regions. Usually these have close relatives that demonstrate much greater species diversity in tropical and subtropical regions. There has been debate about whether the New Zealand taxa are relics of a once more diverse (sub)tropical flora that has largely disappeared due to cooling climate; or outliers that have only recently managed to extend their range south to the warmer northern parts of New Zealand.

The only certain way to establish which option is correct is to look at the fossil record, patchy though it may be. Pollen offers some clues, but wood, leaf and/or flower and fruit macrofossils are the most reliable pointers to past distributions, as they prove the plants were growing onsite at the time of fossilisation. The macrofossils of several taxa at Foulden, including a species of Meliaceae, provide the first and in some cases the only reliable evidence for their definite place as members of New Zealand's paleoflora.

Dysoxylum lies within the large tree family Meliaceae (50 genera and about 650 species). These trees and shrubs are important components of tropical and subtropical rainforests and include globally significant timber trees such as mahogany (*Swietenia* spp.) and red cedar (*Toona ciliata*). These commonly have fairly large, alternate, pinnately compound leaves (with smaller leaflets either side of the central stem), and flowers with prominent staminal tubes and reduced petals. Their fruits are either berries or capsules with fleshy arillate (covered) or dry-winged seeds.[35]

The sole living New Zealand species, kohekohe (*Dysoxylum spectabile*), sometimes called New Zealand mahogany, is a small tree (up to 20m tall) with smooth-margined glossy leaves, sweet-scented nectar-rich flowers, and green leathery capsules that split open to reveal three large, bright orange to red seeds. These are eaten and dispersed by large

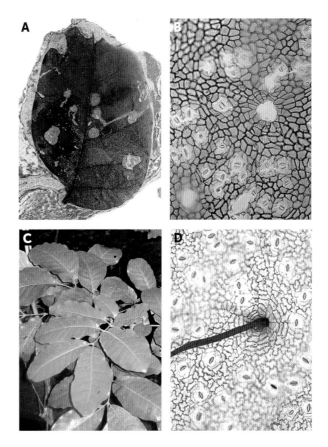

A|B An asymmetrical leaflet and lower cuticle with hair base of '*Dysoxylum*' from Foulden alongside a modern *D. spectabile* leaf (**C**) and cuticle (**D**) for comparison.
JMB, JGC

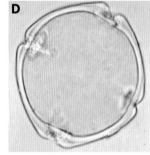

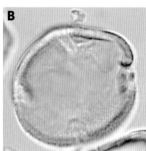

A The fossil '*Dysoxylum*' flower is of the 'dish-bowl' type, with four or five perianth segments, a central, cylindrical, toothed staminal tube and 8–10 anthers included in the tube near the mouth. Pollen of *Tetracolporites spectabilis* was extracted from the anthers (**B**). A flower (**C**) and pollen (**D**) of *D. spectabile* are shown for comparison. UK, JMB, JGC

native birds such as tūī, kōkako and kererū, as well as several introduced species of birds and mammals.[36]

Despite their close resemblance to *Dysoxylum*, placing the fossils into a genus is somewhat problematic. Meliaceae are a large family and the relationships between the genera are complex. Recent advances in molecular phylogenetics have shown that *Dysoxylum* and several large genera previously thought to be separate entities are in fact variously nested within each other: disentangling the resulting taxonomic confusion will not be simple.[37] In the case of the 23-million-year-old fossils, where DNA is not available, other avenues such as morphological comparisons with nearest living relatives and the use of all available lines of evidence become necessary.

It is likely that this species is a distant relative or possible ancestor of the *Dysoxylum* species still living in New Zealand today, but while investigation continues and pending formal naming, the Foulden

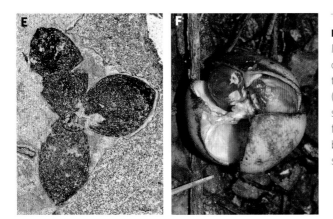

fossil is being labelled '*Dysoxylum*'. As with the living New Zealand species, the fossil plant was probably pollinated by small insects and the fruit dispersed by birds.[38]

MENISPERMACEAE – KIWI CURARE?

Sometimes called moonseeds for the crescent-shaped seeds in some species, Menispermaceae are a family of subtropical to tropical vines with 68 genera and about 440 species.[39] The family is probably best known as one of the sources for the South American poison-dart toxin, curare.

Menisperms have distinctive small, hard, seed-like fruits (endocarps) that are usually highly sculpted, and in some genera look remarkably like little coiled ammonites (an extinct group of shelled marine molluscs). These fruits are fleshy with a single seed and are thought to be dispersed by birds.[40]

Although no members of this family now live in New Zealand, fossil endocarps and leaves with affinities to Menisperms have been found at Foulden and at other fossil localities around the country, in particular at some sites of Pliocene age around Auckland. The endocarps most likely belong to *Hypserpa*, a small genus of rainforest vine with long leaves that is found today in Southeast Asia, Australia and some Pacific islands.

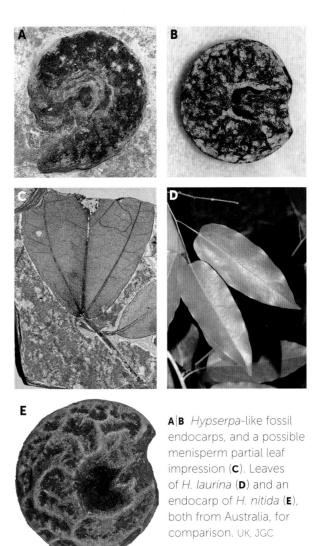

A|B *Hypserpa*-like fossil endocarps, and a possible menisperm partial leaf impression (**C**). Leaves of *H. laurina* (**D**) and an endocarp of *H. nitida* (**E**), both from Australia, for comparison. UK, JGC

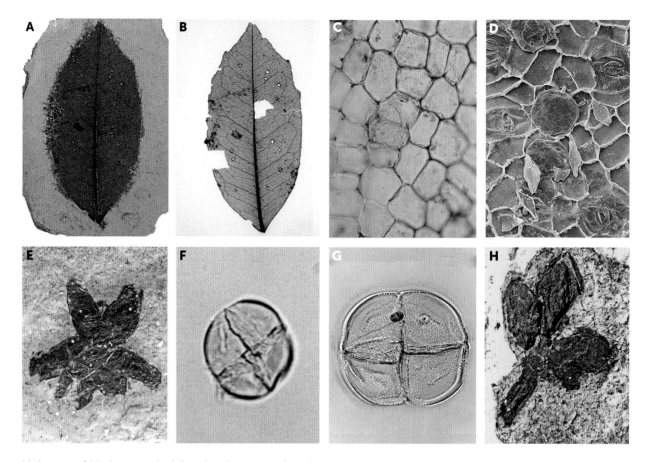

Holotype of *Hedycarya pluvisilva* showing the leaf as it looked after collection in the field (**A**), and the leaf prepared for study (**B**); note the looped veins, small teeth and evidence of insect damage. Upper (**C**) and lower (**D**) cuticle details of *H. pluvisilva* samples seen with a light microscope and SEM respectively.
E One of the dark-coloured fossil *Hedycarya* flowers.
F A pollen grain of *Planarpollenites fragilis*, named for its delicate structure, from an anther on the fossil, forms a distinctive tetrad. **G** A comparative pollen tetrad of modern *H. arborea*. **H** A compressed fossil with three fruits that match the spindle-shaped drupelets of many modern species in the genus.
JMB, EG

MONIMIACEAE – *HEDYCARYA*

Hedycarya is one of the best examples from Foulden of a plant that is represented by leaves with cuticular detail, flowers with in situ pollen and fruit. It is in the basal angiosperm family Monimiaceae, a mainly Southern Hemisphere family of tropical/subtropical trees and shrubs.

Australia has eight genera and 40 species, and New Caledonia has two genera and about 10 species.[41] Modern New Zealand has only a single genus and species, *Hedycarya arborea*, known as pigeonwood or porokaiwhiri, a small, shade-tolerant understorey tree in rainforests in the North Island and in lowland areas of the South Island. The paired leaves are elliptical in shape, up to 10cm long and have toothed margins. The flowers have multiple parts.[42]

Several leaves from Foulden are close matches, although they also resemble one of the two Australian rainforest species. The Foulden leaf fossils have been given the name *H. pluvisilva* (from the Latin *pluviam* for rain and *silva* for forest).

Several small, cup-shaped male flowers were found to contain very fragile pollen that closely resembles that of *H. arborea*, although the grains are much smaller. The flowers were probably insect-pollinated (some modern species are pollinated by thrips and probably also by other small insects).[43] The delicate pollen is unlikely to be preserved in the fossil record except in very low-energy and anoxic depositional environments such as a maar.

One flattened, blackish fruit made up of three drupelets also closely resembles that of modern *H. arborea*. In New Zealand forests today the fruits may be red, orange, yellow or black, and are eaten and dispersed readily by fruit-eating birds such as kererū, hence the common name.[44] The presence of these and other large fleshy fruits is another indication that fruit-eating birds were part of the local fauna.

Bird dispersal also helps explain the biogeographic distribution of the Monimiaceae, as the family has a long – if patchy – fossil history across the landmasses that were once part of Gondwana.[45] It seems likely that it was once widespread across the southern continents, including probably Antarctica.

I The modern New Zealand *Hedycarya arborea* for comparison, showing a cleared leaf (chemically treated to make the leaf transparent). **J** Lower cuticle under SEM. **K** Male flower (with a tiny pollinating thrips insect) and **L** infructescence with fusiform drupelets (note also the leaf with insect damage). JGC

MYRTACEAE

One of the major flowering plant families of the Southern Hemisphere, Myrtaceae – the myrtle family – comprise 132 genera and about 6000 species worldwide, with centres of diversity in Australia and South America.[46] In New Zealand there are five native genera and 27 indigenous species, 26 of which are endemic. These include the iconic rātā and pōhutukawa (both species of *Metrosideros*), mānuka (*Leptospermum*) and kānuka (*Kunzea*). The family is known from fossil leaves and fruits in the Eocene and Miocene of New Zealand.[47]

Although some of the fossil leaves at Foulden belong to the Myrtaceae, the most abundant fossils for the family are seeds – small globular fossils, often partly replaced by pyrite, which are the remains of fleshy fruits related to those of the modern maire (tawake or *Syzygium maire*), a small tree of swampy areas. These are common in the deposit, and dozens are sometimes preserved on a single bedding plane, suggesting local seasonal mass fruiting near the lake edge.

Although rātā species are now the dominant native trees in many areas of New Zealand, providing spectacular summer floral displays, Miocene-aged fossils of this genus are surprisingly rare at Foulden and elsewhere.[48] Several leaves at the site possess *Metrosideros*-like venation and are covered in the familiar densely spaced oil glands, but as they lack cuticle their identity cannot be confirmed. In contrast, there are several ribbed capsular fruits at Foulden which, although they resemble the genus, are structurally much closer to members of the largely New Caledonian *Metrosideros* subgenus *Mearnsia* than to living New Zealand taxa.

The fruits of *Metrosideros* release numerous tiny seeds that are capable of both wind and oceanic water dispersal. Some 50–60 species can be found throughout the Pacific Islands and as far away as Hawai`i and South America.[49]

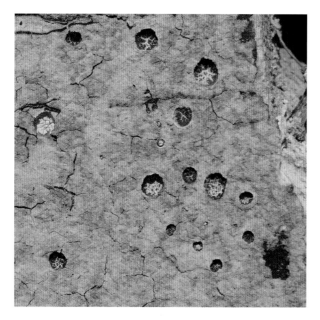

Some bedding planes are dotted with dozens of small globular holes, which represent the remains of fleshy *Syzygium*-like Myrtaceae single-seeded fruits. JKL

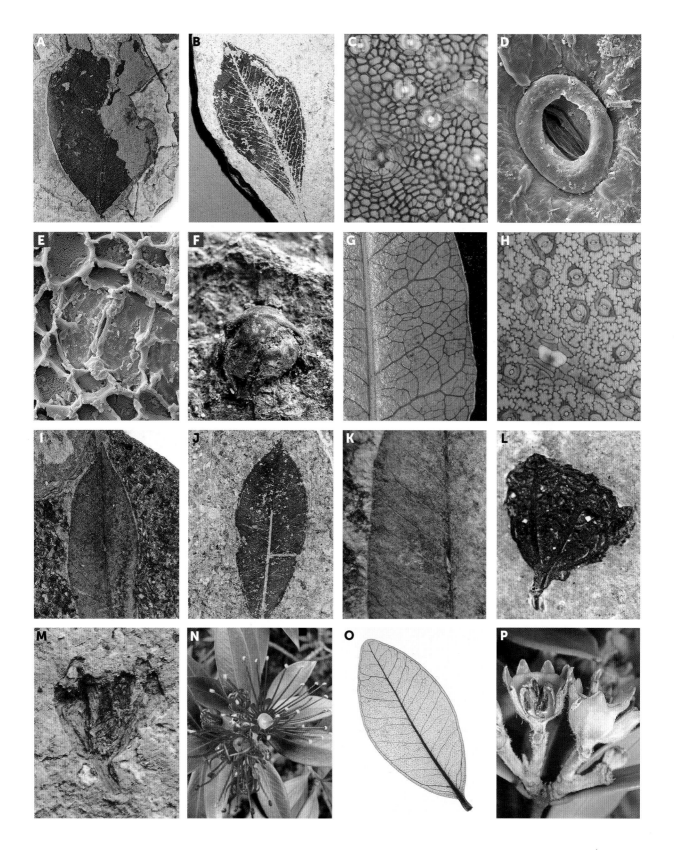

A

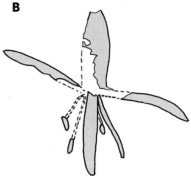

B

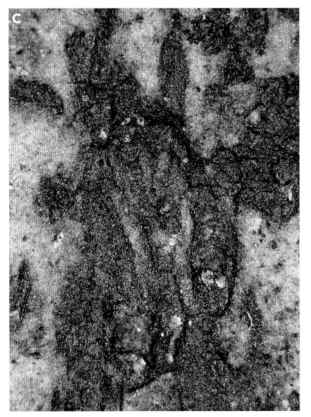

C

ONAGRACEAE – *FUCHSIA* AND BIRD-POLLINATION

One flower was a puzzle – until we upended the image
and saw the resemblance to the pendant, tubular bell-
shaped *Fuchsia* flower. As it lacked pollen, however, no
further identification was possible. Later, some odd-
looking brownish plant fragments were discovered
resting on a small twig. When prepared, the fragments
proved to be detached paired anthers from which we
squeezed out distinctive diporate (two-pored) pollen
grains covered with viscin threads that help to tie
the pollen grains into masses. Such pollen grains are
produced by *Fuchsia*, and similar dispersed grains are
present in the New Zealand fossil record as far back
as the Oligocene. The sticky tangles of pollen grains
ensure that a pollinator visiting a flower takes extra
pollen away with it. They are common in the Foulden
Maar core.

These are the first fossil *Fuchsia* flower and anthers
from any fossil site worldwide. Since the parent plants

must have been living on the lake edge, the absence of leaves is probably due to the fragility of both leaves and cuticle, suggesting that the deciduous habit of one of the living New Zealand species may have been already established by the late Oligocene.

Fuchsia today has an intriguing biogeographic pattern. There is one small tree endemic to Tahiti; and three endemic New Zealand species: kōtukutuku or tree fuchsia (*Fuchsia excorticata*), a familiar and widespread small deciduous tree with papery peeling bark and, at up to 12m tall, the largest species in the genus; *F. perscandens*, a widespread but uncommon scrambling shrubby vine; and *F. procumbens* or creeping fuchsia, a prostrate groundcover restricted to warmer coastal areas of the North Island. The remaining 100 plus *Fuchsia* species, from which the New Zealand and Tahitian species diverged around 30 Ma occur in South and Central America as far north as Mexico, where they are mainly pollinated by hummingbirds and other nectar-feeding birds.[50]

The New Zealand species are pollinated by a range of native and introduced birds, especially honeyeaters such as tūī and korimako (New Zealand bellbird), both of which can often be seen with a dusting of the conspicuous bright blue pollen of the native *Fuchsia* species on their feathers.[51] The bones of mid-Miocene-aged honeyeaters are present in the St Bathans fauna. Although these are several million years younger than the Foulden site, their ancestors may have been among the pollinators and dispersers in the forest around Foulden.

How did *Fuchsia* get from earliest Miocene New Zealand to Tahiti and South America (or vice versa)? The answer is probably with birds, since the fleshy, seed-filled berries (kōnini) are palatable to birds, including kererū.[52] Before the arrival of humans, large fruit-eating pigeons were abundant on many Pacific islands, including Tahiti, and these may have transported the seeds of that species' ancestor there from New Zealand.[53] But until we know more about the fossil history of *Fuchsia* in America, we can only speculate on the direction of dispersal.

Pollen grains of *Koninidites aspis* extracted from the *Fuchsia antiqua* anther mass, viewed under light microscopy (**D**) and SEM (**E**); note the viscin threads that tie the pollen grains together. Flowers of the modern New Zealand tree fuchsia (*F. excorticata*) for comparison, showing the colour change (**F**) between young and older flowers. **G** A dissected anther mass from an unopened bud. JMB, EG, JGC, UK

A|B *Myrsine* leaves from Foulden Maar, one showing leaf galls (as seen on many of the fossils for this taxon).

PRIMULACEAE (MYRSINACEAE)

The genus *Myrsine* has 11 indigenous New Zealand tree and shrub species, collectively called matipo or māpou. Although traditionally these have been placed in the family Myrsinaceae (muttonwoods), recent molecular studies have shown that they are nested instead within the worldwide and now rather diverse primrose family, Primulaceae.[54]

The *Myrsine* specimens at Foulden are preserved most commonly as leaves, often with numerous insect galls. *Myrsine* leaves have distinctive cuticle with prominent hair bases, as well as resin glands.[55] Several fossils have been reported previously from Miocene New Zealand, including from Foulden Maar.[56]

Wood and bark are rare at Foulden; only a few twigs and bits of larger branches have been seen to date, which is surprising given that lush forest grew right around the lake shore. The fossilised *Fuchsia* anther mass described above was discovered sitting on a small twig, but examination of the twig's wood anatomy showed that instead of being from *Fuchsia*, as we had hoped, it appeared to be much closer to *Myrsine*.[57] Unfortunately, it was too small to be definitive.

PROTEACEAE

Many plant families are represented at Foulden by both leaves and dispersed pollen, including the Proteaceae, a largely Southern Hemisphere family of evergreen flowering trees and shrubs, often with brightly coloured blossoms.[58] Members of this family tend to occupy one of two very different habitats: some grow in rainforests, but the majority occupy drier regions with poor soils.

Many Australian species are adapted to fire. *Banksia*, for example, has woody fruits that release their seeds after burning, and some other species resprout following fire. Most of the family also have unusual clustered roots that are adapted to taking up nutrients from impoverished soils.[59]

Most Proteaceae are found today in South Africa, South America, Australia and New Caledonia. There

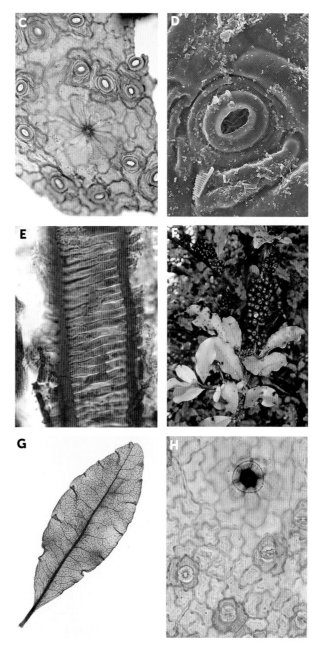

These leaves show characteristic stomatal (**C**) and hair base-pattern (**D**) for the genus. **E** Vessels in a small piece of fossil Primulaceae-like wood similar to *Myrsine* (stained red with Safranin O). **F|G** Modern *Myrsine australis* leaves and cuticle (**H**) for comparison. JMB, JGC

are just two native species in the New Zealand flora. One of these, rewarewa (*Knightia excelsa*), is a forest tree that is widespread in warmer northern areas and extends as far south as the Marlborough Sounds; toru (*Toronia toru*) occurs in the North Island as far south as Lake Taupō.[60]

There is an excellent fossil record for Proteaceae in Australasia, and New Zealand pollen records extend back to the Cretaceous. Fossilised cuticle belonging to the genera *Helicia*, *Macadamia*, *Musgravea* and tribes Gevuineae (cf. *Hicksbeachia*) and Embothrieae first appears here in the Paleocene.[61] Fossil Proteaceae leaf taxa described from New Zealand also include extinct examples of the Australian genus *Banksia*, a *Persoonia*-like possible *Toronia* relative, and others with affinities to the tribes Proteoideae and the New Caledonian endemic genus *Beauprea*.[62]

In addition to reports of dispersed Proteaceae cuticle at Foulden, two new species of Proteaceae have been described.[63] Each is based on a single leaf only, but the leaf characters, and especially their distinctive cuticles, make it possible to assign them to different subtribes within the family that are no longer present in New Zealand. *Euproteaciphyllum alloxynoides* is a

---

A|B|C|D With leaves 25cm long and 13cm across, *Euproteaciphyllum alloxynoides* is most closely related to a low-latitude rainforest species in the genus *Alloxylon* that grows in northern Queensland, Papua New Guinea and New Caledonia. UK, JGC

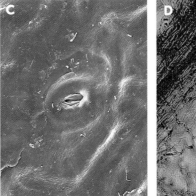 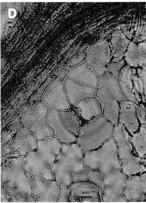

E|F *Euproteaciphyllum pacificum* holotype leaf (or leaflet) and cuticle under SEM (**G**) showing stomates and a characteristic Gevuineae-type hair base. JMB, EG

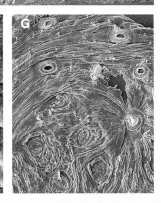

large multi-lobed but untoothed leaf showing affinities to the eastern Australian genus *Alloxylon* (Queensland tree waratah), while the unlobed but toothed leaf (or possibly isolated leaflet) of *E. pacificum* is related to the tribe Gevuineae, found today in Australia, the Pacific and South America.

Modern examples of (**H**) *Alloxylon flammeum* deeply lobed juvenile foliage; (**I**) flowering *A. pinnatum*; **J|K** pressed *Gevuina avellana* showing compound leaves with toothed leaflets and close-up of an apical leaflet for comparison. CASLIBER (CC-BY-3.0) https://commons.wikimedia.org/wiki/, CFile:Alloxylon flammeumimm2, JGC

FUNGI

Fungi are not plants but are still key elements of rainforest ecosystems, where they help to recycle nutrients. They are ubiquitous in modern New Zealand forest ecosystems: as saprophytes feeding on decaying material; as pathogens; as mycorrhizae living in symbiosis with plant roots; and as epiphytes living harmlessly on the surface of leaves. With more than 8000 species (some estimates suggest up to 25,000) – about 25 percent of them endemic – fungi form a major component of the modern New Zealand biota. However, since they typically leave few traces in the fossil record other than dispersed spores, the fossil record of fungi in New Zealand is almost non-existent.[64]

Leaf surfaces provide habitats for a wide range of organisms, including bacteria, fungi, algae and mosses. Subtropical rainforest trees are mainly evergreen and their leaves remain attached for at least 18 months and up to five years, which allows time for organisms such as fungi to colonise and complete their life cycles. Evidence of these organisms can sometimes be found on fossil leaves, especially in warm, wet paleoenvironments.[65]

Although fungi are rare on leaves at Foulden, they are quite diverse. Species of *Entopeltacites* have been found on six leaves. This *Vizella*-like fungus sits below the cuticle and, as such, is protected from drying out. *Vizella* does not require high-humidity conditions and can be found today on leaves in Dunedin. Other genera of fungi, including *Callimothallus*, *Trichopeltinites* and several unidentified fungal shields and perithecia (fruiting bodies), have also been found at Foulden, but these are rare. The only other epiphyllous (growing on leaves) forms found were a small number of fungal germinating spores of a species of *Desmidiospora*. In contrast, leaves from a late Eocene rainforest site in Southland had an abundance of many forms of epiphyllous fungi, indicating higher rainfall and warmer, more humid conditions. It seems likely that at the time the Foulden leaves were growing, there were at least seasonally dry periods.[66]

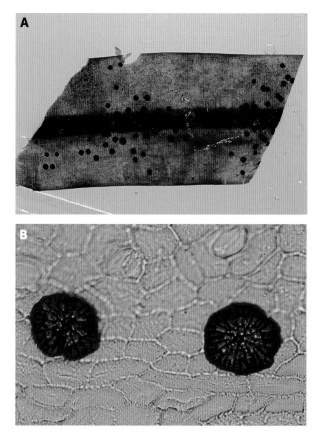

Fossilised fungi: Epiphyllous fungi on *Podocarpus travisiae* leaf fragment (**A**), and two *Callimothallus* shields (**B**).

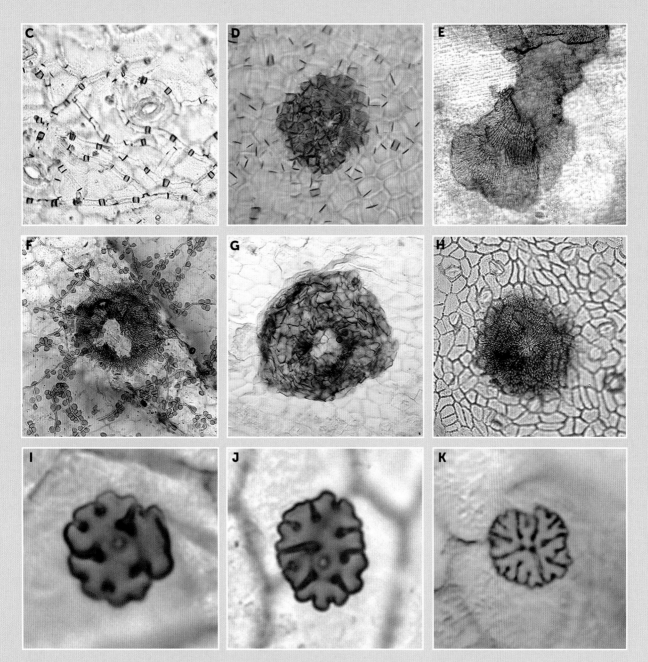

Entopeltacites hyphae (the thread-like branching filaments that comprise the body of the fungus) (**C**), fruiting body (**D**) and *Trichopeltinites* (**E**).
F|G Two unidentified perithecia (hollow, fungal fruiting bodies that open by apical pores); and (**H**), a microthyriaceous shield (disk-shaped, flat fruiting body).
I|J|K Fungal germlings, indicating warm wet conditions. JMB

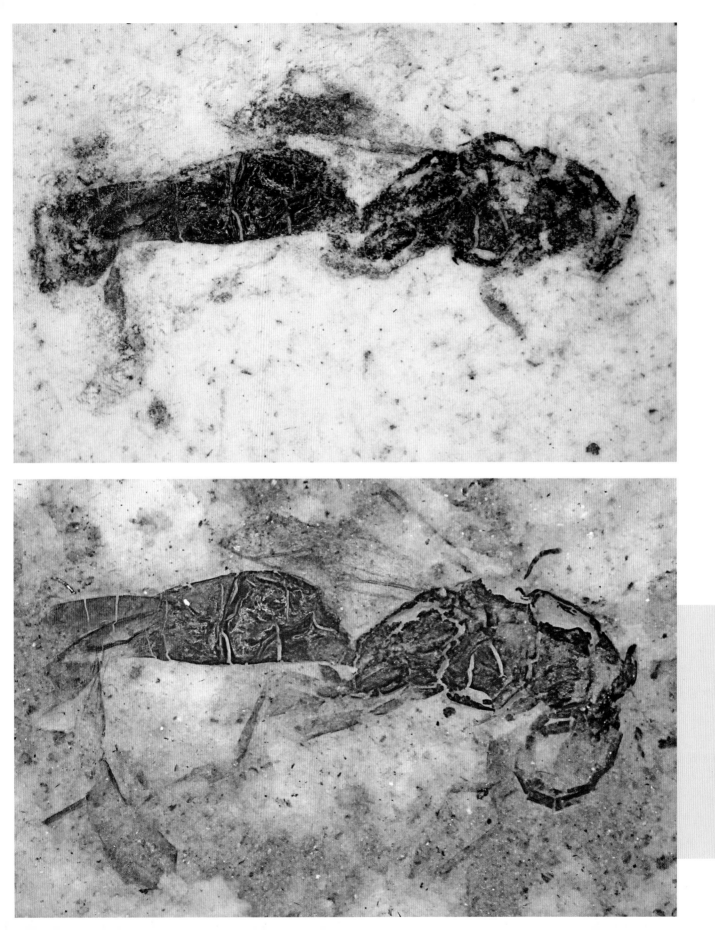

CHAPTER 6
SPIDERS AND INSECTS

Fossil insects and spiders from Foulden Maar are completely transforming our knowledge of the history of terrestrial invertebrates in New Zealand. About half the genera represented in the fossil discoveries are locally or globally extinct. All specimens represent first fossil records from New Zealand, and often also for the Southern Hemisphere or even globally.

New Zealand today has a rich and diverse fauna of land-dwelling arthropods (invertebrates with jointed limbs, a segmented body and an exoskeleton) – mostly insects and spiders. It includes about 14,000 species of insects, 90 percent of which are endemic. There are more than 1700 spider species, many of which are yet to be described.[1]

But until investigations began at Foulden Maar in 2003, only six fossil insects more than two million years old had been found in New Zealand. As well as two Eocene fossils, these included a Triassic beetle elytron (wingcase), a Jurassic orthopteran wing (orthoptera are grasshoppers, locusts, crickets and their relatives) and two Cretaceous fossils: a moth sclerite (hardened body part) and another beetle elytron. Four of these date back to the time when Zealandia was still part of Gondwana: only two were true locals.

The first fossil insects at Foulden were found in 2006 by Jennifer Bannister, who discovered a large leaf with scale insects attached (see page 126).

OPPOSITE A fossil wasp from Foulden Maar before (top) and after preparation. Preparation, photographing under ethanol and digital image stacking have revealed fine details of the insect's antennae, legs and wings. UK

RECORDING THE FIND

Preparing and photographing tiny fossils is painstaking work. The small blocks containing both sides of the insect (part and counterpart) are cut from the diatomite with a knife and placed in plastic bags to keep them moist.

In the lab, the insect is carefully excavated, cleaned with fine needles and paintbrushes and photographed using a camera attached to a stereomicroscope, sometimes under ethanol to accentuate fine details. Photomicrographs are taken from up to 30 individual focal planes to compensate for the uneven surface of the fossil and the diatomite matrix, and these are then digitally merged into a composite image.

Drawings made from these photomicrographic composites are checked against the original specimen. The tiny morphological features are measured using an ocular micrometer (a lens with a ruler engraved on it).

Insects are then stored in sample bags in a refrigerator in order to prevent desiccation.

PROPORTION OF INSECT ORDERS IN THE
FOSSIL FAUNA AT FOULDEN

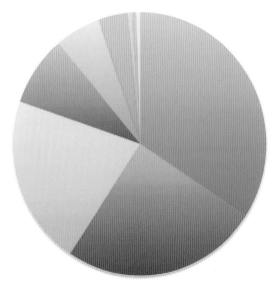

- Coleoptera (beetles)
- Diptera (flies)
- Hymenoptera (wasps, ants)
- Isoptera (termites)
- Hemiptera (true bugs)
- Trichoptera (caddisflies)
- Mecoptera (scorpionflies)
- Megaloptera (alderflies)
- Odonata (dragonflies, damselflies)

Leaf and fish fossils are often easily recognised in the diatomite, but finding the remains of insects and spiders requires a keen eye, a hand lens and an abundance of patience, and many hours or even days may go by without a single one being found.

Insect fossils are randomly distributed throughout the exposed 10m section of diatomite at Foulden and are generally found in the light-coloured layers deposited in the spring and summer. The dark organic layers are virtually free of insects and spiders. About half of the finds are more or less complete specimens, often only fully revealed after further work; and the rest are isolated fragments such as wings or other body parts.

The most abundant insects at Foulden are from the orders Coleoptera (beetles), Hymenoptera (ants, bees and wasps) and Diptera (flies). Together with Lepidoptera (butterflies and moths), these are by far the most diverse insect orders globally. Their evolutionary success is based on a complete metamorphosis (holometabolism) that includes four life stages: egg, larva, pupa and adult. Interestingly, no Lepidoptera fossils have been found at Foulden Maar, which is almost certainly due to 'preservation bias' – the fact that their wings are made of tiny scales that quickly disintegrate on death – rather than an indication of their absence in the early Miocene.

At least 17 genera from nine insect orders have been found.[2] About half of them are extinct (locally or globally), while the other genera still exist in New Zealand today. Seventy-two percent of the specimens are terrestrial adults from the rainforest surrounding the maar, and 28 percent are fully or partly aquatic taxa, mainly preserved as pupae that lived in the maar lake.

Except for two flat bugs of the same genus, each specimen represents a different kind of insect, telling its own story and contributing towards our knowledge of the insect diversity in New Zealand 23 million years ago. There is much work still to be done, as thousands of insects remain buried in the Foulden diatomite.

This chapter discusses the six most common insect orders found at Foulden. Odonata (dragonflies and damselflies), Megaloptera (alderflies and dobsonflies) and Mecoptera (scorpionflies) are not included because, although each is represented by one specimen, they have yet to be described. All are first fossil records from New Zealand.

SPIDERS – ORDER ARANEAE

Spiders belong in the same class as mites, ticks, harvestmen and pseudoscorpions. They differ from insects in that they have eight legs rather than six, two body parts instead of three, and they lack antennae and wings.

Spiders are usually the top terrestrial invertebrate predators in New Zealand forest ecosystems, and use venom to subdue their prey. All types produce spider silk for webs and to aid travel.

In the modern New Zealand fauna there are over 1600 known endemic spider species (95 percent of the total) assigned to 137 genera or subgenera. Globally, there are 120 spider families known; New Zealand has representatives from 57, including one family that is endemic.[3] Until the discovery of spiders at Foulden, nothing at all was known of their fossil history in New Zealand.

A possible male orb weaver spider, showing its elongate, slender legs. The pyritised fossil cannot be assigned to a genus with certainty. UK

Of the four spiders found to date, two are preserved as pyrite, which covers or replaces the original cuticle and thus obscures most anatomical details. One of these may be a spider or a harvestman and is referred to as Arachnida *incertae sedis*. The other is a spider also of uncertain placement (Araneomorphae *incertae sedis*), possibly in the family Tetragnathidae (long-jawed orb weavers), which has several endemic and introduced genera throughout New Zealand.[4]

The two non-pyritised specimens belong to the Mygalomorphae, a group of spiders with robust bodies, stout legs, strong jaws, and fangs that inject venom into their prey. Among these powerful predators are the tarantulas and the Australian funnel-web spiders. One of the fossils is an adult male (based on its complex pedipalps, the appendages in front of the legs) but is incompletely preserved and is referred to as Mygalomorphae *incertae sedis*.

The second mygalomorph is a well-preserved complete specimen with a body length of 5.5mm and a total length (including the legs) of at least 11mm. This fossil is placed in Idiopidae (armoured trapdoor spiders), a family with a mainly Southern Hemisphere distribution but extending as far north as Central America, Morocco and India. The only New Zealand genus is *Cantuaria*, of which 42 species are distributed predominantly in grasslands of the South Island. The specimen from Foulden is a juvenile that may have fallen into the maar lake when leaving its burrow; alternatively, its burrow may have been washed into the lake during heavy rain.

The fossil spiders from Foulden, although relatively poorly preserved and few in number, provide evidence for a possibly diverse spider fauna in New Zealand's warm forests about 23 Ma. Spider fragments and spider silk, in some cases with prey attached, recently discovered in early Miocene (19–16 Ma) amber (fossilised tree resin) from Central Otago, are yet to be described.

A The fossil of a juvenile trapdoor spider whose features place it in the family Idiopidae (armoured trapdoor spiders). **B** An extant Idiopidae for comparison. UK, Victoria Wood, Paul Selden

TERMITES – ORDER BLATTODEA

Findings at Foulden have similarly greatly extended our understanding of New Zealand's early termite fauna. Termites are eusocial insects (existing in cooperative groups) in the infraorder Isoptera, now sometimes included in Blattodea, along with cockroaches, to which they are closely related. More than 3100 species of termites are known to science. They are ecologically important in breaking down wood, plant matter and dung in tropical and subtropical regions. New Zealand has a small termite fauna of only nine species, six of them introduced from Australia, which has a much more diverse termite fauna of 360 described species.[5]

Of the three endemic New Zealand species, *Stolotermes ruficeps* and *S. inopus* are dampwood termites (Stolotermitidae), and *Kalotermes brouni* is a drywood termite (Kalotermitidae). These occur as small colonies of up to several hundred individuals in indigenous or exotic wood, predominantly in the North Island.

Fossil termites are relatively common and diverse at Foulden Maar. Their presence was not surprising since fossilised termite faecal pellets (frass) had been described from Miocene silicified wood near Kaipara.[6] They are the only termite body fossils from New Zealand, and they provide unique insights into the

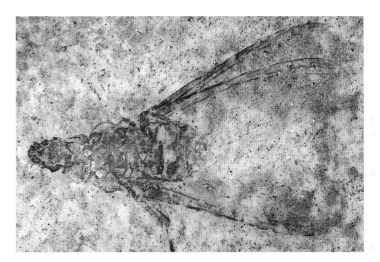

The fossil of a winged reproductive termite (alate) in ventral preservation, meaning its underside is exposed.
UK

'NEW' ANCIENT SPECIES NAMED

Three of the new termite genera and two species from Foulden are named after the type locality Foulden Maar (*Pterotermopsis fouldenica*) or the wider geographic area in which they were discovered: Waipīata (*Waipiatatermes*), Taieri River (*Taieritermes*), Otago (*Otagotermes*) and New Zealand (*Otagotermes novazealandicus*). Another species, *Stolotermes kupe*, is named after Kupe, an early navigator/tohungamoana who came to Aotearoa from the Pacific homeland of Hawaiki.

GENUS AND SPECIES	FAMILY	HABITAT
Pterotermopsis fouldenica	Kalotermitidae	drywood termite
Waipiatatermes matatoka	family uncertain	–
Taieritermes krishnai	Kalotermitidae	drywood termite
Otagotermes novazealandicus	Kalotermitidae	drywood termite
Stolotermes kupe	Stolotermitidae	dampwood termite
Genus and species indet.	family uncertain	–

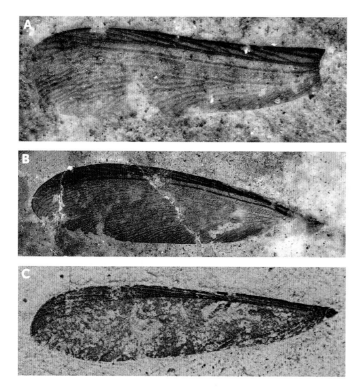

A An isolated alate wing of *Taieritermes krishnai* (Kalotermitidae). The species' name honours the late Kumar Krishna (1926–2014), world authority on the Isoptera.

B An isolated alate wing of *Waipiatatermes matatoka*. Mātātoka is the Māori word for fossil.

C An alate wing of the dampwood termite *Stolotermes kupe* (Stolotermitidae). Two living species of *Stolotermes* occur in New Zealand, both endemic. UK

early Miocene diversity and the biogeography of this group.

Four genera and six species of termites have been described from 16 Foulden fossils.[7] Of these, only the genus *Stolotermes* persists in the extant fauna, whereas the various kalotermitid genera have become extinct. Together these species demonstrate that rainforest at Foulden Maar in the early Miocene supported a much higher diversity of termites than modern ecosystems in New Zealand. Climate cooling in the late Miocene and during Pleistocene glacial cycles is the most likely cause for the decline.

Most of the 16 specimens (8 percent of all identified insects) from Foulden are alates or their wings. Alates are winged females and males that pair up during nuptial flights and shed their wings shortly afterwards. Because they are poor flyers and their wings are easily dispersed by wind, alates are much more likely to be fossilised in lake sediments than the forest-dwelling worker and soldier termites.

With the exception of a kalotermitid-like wing of uncertain family placement, all termites from Foulden belong to the same two families as the existing New Zealand species: Kalotermitidae and Stolotermitidae. Both are ancient lineages that evolved in the Mesozoic when Zealandia was part of Gondwana. More 'modern' termites (Neoisoptera), which diversified mainly during the Cenozoic after Zealandia separated from Gondwana, are absent from Foulden. This may reflect a long isolation and independent evolution of termites in New Zealand, with the early Miocene species from Foulden representing the last remnants of now extinct ancient groups of Kalotermitidae.

BUGS – ORDER HEMIPTERA

The order Hemiptera (meaning 'half-winged') includes over 80,000 species in groups such as cicadas, aphids, scale insects, planthoppers, leafhoppers and shield bugs. The name of the order refers to the structure of the forewing, which has a thickened leathery base and a membranous tip.

Hemiptera are highly variable in body form, size (under 1mm to 150mm) and behavioural features, and occur in a wide range of environments. Most are terrestrial, but there are aquatic groups such as water striders. A distinctive character of all hemipterans is the structure of their piercing and sucking mouthparts, which are adapted for extracting plant juices, or liquids from animal prey.

Today there are some 1100 bug species in New Zealand, including many endemics.[8]

Almost 6 percent of the Foulden insect fossils found so far belong to the Hemiptera, including representatives of at least five families. At the time of their discovery, these were the only Hemiptera fossils in New Zealand. Others have subsequently been discovered at Oligocene and Miocene sites in Otago, including a planthopper nymph and male scale insects in amber, an extinct tribe of the planthopper family Tropiduchidae from the Manuherikia Group near Bannockburn, and an extinct genus and species of primitive cicada (Tettigarctidae) from Hindon Maar.[9]

A rare worker or soldier termite found at Foulden. The missing parts of the specimen are preserved on the opposite side of the diatomite slab (the counterpart). UK

SCALE INSECTS

Fourteen small pale dots, discovered on an angiosperm leaf (possibly Elaeocarpaceae) by Jennifer Bannister in 2006 and later identified by Tony Harris as female scale insects, were the first fossil insects described from Foulden Maar.[10] The discovery of female scale insects in situ on fossil leaves provides an extremely rare example of a 23-million-year-old insect–plant association.

Belonging to the family Diaspididae (with armoured scales), they consist of nine mature females, and five smaller females that had not long moulted into their third and final developmental stage.

The term 'armoured scales' is a reference to the waxy shield that helps protect scale insects from predators. Still waxy and nacreous as they were in life, these fossilised scale insects were found attached to the leaf veins from which they extracted sap 23 million years ago. They have no close relatives in New Zealand today and probably went extinct, together with the host tree, in response to climate cooling.

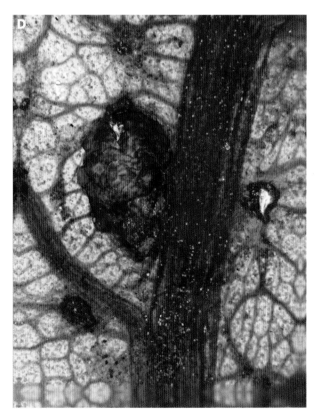

A|B|C A line drawing and photographs of the 14 female scale insects found attached along the veins of a fossil leaf. The waxy scales are about 2.5mm in size. Remarkably, they have retained their nacreous colouration. JMB, Anthony Harris (Otago Museum)

D 'Pink frilly', a second type of female scale insect, possibly a mealy bug (Coccidae), attached to a fossil *Cryptocarya*-like (Lauraceae) leaf. JMB

A second type of scale insect at Foulden, referred to informally as 'pink frilly', is quite different in appearance and may belong to the family Coccidae (mealy bugs), which includes soft scales. This scale was found still attached to a *Cryptocarya*-like leaf, an evergreen laurel genus that is now extinct in New Zealand but still common in Australia and Asia, with a fossil record going back to the Cretaceous.

Scale insects are cosmopolitan plant parasites, more common in subtropical and tropical environments. Although there are about 28 families and more than 7000 living species worldwide, as fossils they are rare. The 50 species native to New Zealand today are usually found on forest shrubs and trees.

FLAT BUGS

Two insects from Foulden are placed in the family Aradidae, generally known as flat bugs or bark bugs.[11] They represent the first Southern Hemisphere fossil records for the family.

Aradidae include more than 200 genera and 2000 species worldwide, but their fossil record is rather scarce, with only some 40 taxa known. These are mainly from amber deposits of the Northern Hemisphere; the oldest is from mid-Cretaceous (100–90 Ma) Burmese amber.

New Zealand has considerable aradid diversity, with 39 genera from all eight subfamilies present, including many endemic species. About 65 percent of species are flightless. New Zealand Aradidae appear to have a long evolutionary history linked to rainforest habitats, in particular to Gondwanan elements such as podocarps and *Nothofagus* (southern beeches).[12]

Both Aradidae fossils from Foulden are placed in the genus *Aneurus*, based on their small size (5mm), body shape, and characteristics of the head, antennae and abdomen. There are six *Aneurus* species throughout New Zealand today, living under tree bark in lowland to subalpine regions and feeding on the vegetative parts of wood-rotting fungi. At Foulden, *Aneurus* must have lived in trees close to the maar lake, and the two specimens likely drowned when their host trees fell into the lake.

EXTREME GENDER DIFFERENCES

Scale insects are highly sexually dimorphic – that is, males and females exhibit different morphology and behaviour. The immobile females are soft-bodied, wingless and have greatly reduced legs or none at all. Protected from desiccation and predators by a scale cover, they spend their entire life attached to their host plant. In contrast, the mobile adult males have legs, antennae, functional forewings, greatly reduced hindwings and are small and fragile. Lacking functional mouthparts, they live for only a few hours.

A|B Two fossil flat bugs of the genus *Aneurus* represent the first Southern Hemisphere fossil record of the family Aradidae. **C** The extant species *A. zealandensis* for comparison. UK, Birgit Rhode

An incomplete cixiid planthopper fossil that is missing its head, parts of the wings and legs. UK

PRIMITIVE PLANTHOPPERS

Primitive planthoppers are distributed worldwide, with 2000 species in more than 150 genera.

New Zealand is regarded as a biodiversity hotspot for Cixiidae because eight of the 11 cixiid genera recognised are not found elsewhere.[13] They include ancient taxa and many species that are restricted to small, remote areas or offshore islands (e.g. Manawatāwhi Three Kings or Rēkohu Chatham Islands).

At least one fossil insect from Foulden is a small (5mm-long) planthopper. Although it is missing its head and most of its legs, it resembles the living genera *Semo* or *Aka*. Extant species of both genera occur in New Zealand: *Semo* in subalpine shrublands, and *Aka* in beech forests, mixed beech-*Nothofagus*-broadleaf forests and shrublands in lower altitudes.

An isolated wing from Foulden probably represents a second fossil species of cixiid planthopper.

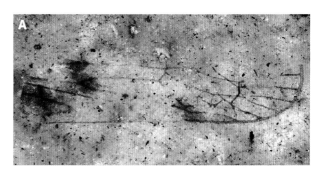

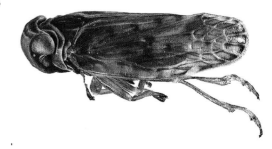

A The isolated wing fragment of a cixiid planthopper.
B Extant species *Semo westlandiae* for comparison.
UK, Birgit Rhode

A LEAFHOPPER WING

One isolated hindwing from Foulden is the first New Zealand fossil record of a leafhopper in the family Cicadellidae. With 78 species in 29 genera in the New Zealand fauna, Cicadellidae is the most diverse family of the hemipteran suborder Auchenorrhyncha living in New Zealand today.[14] Species occur in most types of vegetation, from low-growing plants to shrubs and trees. Except for a few species that consume plant tissue, most New Zealand leafhoppers feed on plant sap. The hindwings are not a typical diagnostic feature in the classification of Cicadellidae, so it is not possible to assign this fossil wing to a subfamily.

A LACE BUG?

An incomplete insect fossil, found lying on its back, preserves remnants of the intricate, lace-like pattern typically present on the pronotum – the tough cuticle covering the first segment of an insect's thorax – and forewings of lace bugs (family Tingidae). Only two species occur in New Zealand today: the widespread endemic Astelia lace bug *Tanybyrsa cumberi* and the introduced rhododendron lace bug *Stephanitis rhododendri*. The common names of both species are derived from the respective host plant genera, which are a food source.

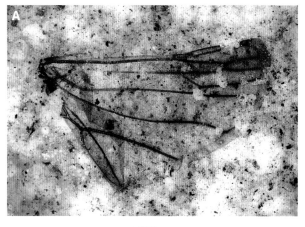

A The hindwing of a fossil leafhopper (Cicadellidae).
B Extant leafhopper (*Novothymbris zealandica*) for comparison. UK, Birgit Rhode

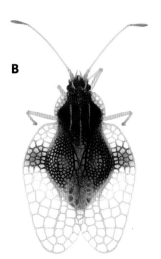

A A probable lacebug (Tingidae) in ventral view, showing remnants of the lace-like wing venation.
B Extant Astelia lacebug *Tanybyrsa cumberi* for comparison. UK, Birgit Rhode

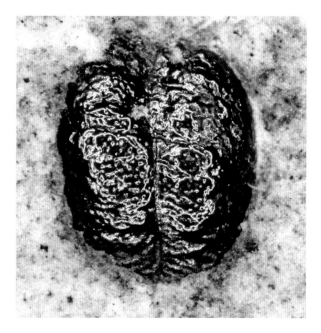

BEETLES — ORDER COLEOPTERA

Beetles are the most common insect fossils at Foulden Maar, accounting for 34 percent of identified specimens.[15]

Coleoptera is an extraordinarily diverse insect order of more than 380,000 known species that populate almost every terrestrial and aquatic environment worldwide. Beetles are readily recognised by their hardened forewings (elytra or wing cases), which protect the hindwings and abdomen. New Zealand has almost 5500 beetle species and more than 100 genera from 82 families today; the estimated total number of species, including those yet to be formally described, is as high as 10,000.[16]

Today's fauna is an assemblage of ancient Gondwana lineages and younger additions that have arrived from elsewhere. Almost all the native species are endemic. Worldwide, beetles have a rich fossil record from the Cretaceous and Cenozoic periods (145 Ma and younger); the oldest records are from the Permian or Jurassic.

Unfortunately, most of the specimens found at Foulden are incomplete, missing crucial diagnostic characteristics, so it is not possible to identify them. Except for a few aquatic larvae, all are adults or their robust elytra, which are particularly prone to fossilisation; some even retain their metallic blue-green structural colour. They are different shapes and sizes (2–23mm long), suggesting that a wide range of families is represented.

The two most common specimens appear to be weevils (Curculionidae) and rove beetles (Staphylinidae), which are the most diverse families in today's New Zealand beetle fauna. Other families present include water scavenger beetles (Hydrophilidae), click beetles (Elateridae), sap beetles (Nitidulidae), longhorn beetles (Cerambycidae) and leaf beetles (Chrysomelidae), all of which have extant species in New Zealand.

ABOVE AND OPPOSITE
Examples of fossil beetles from Foulden Maar. UK

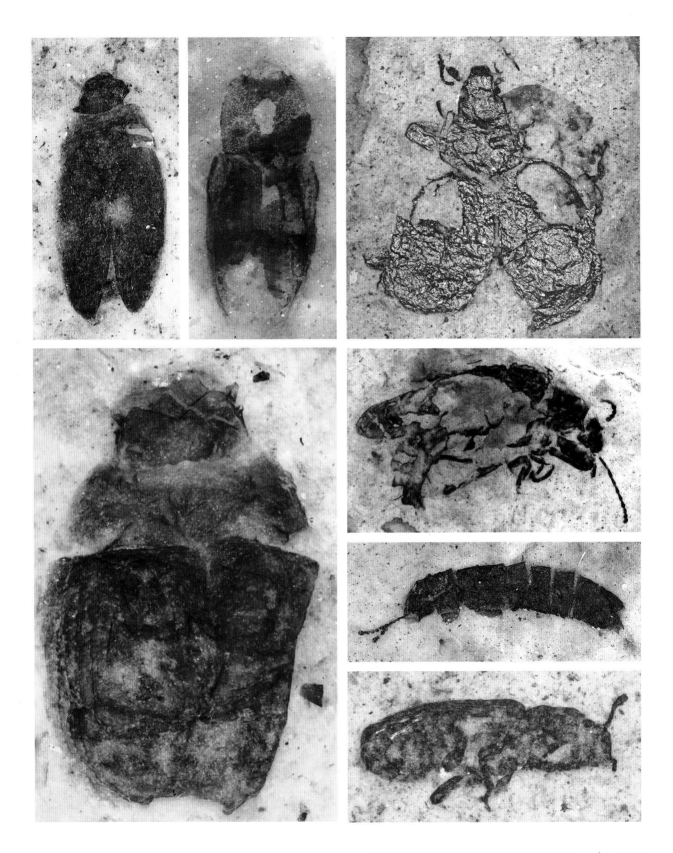

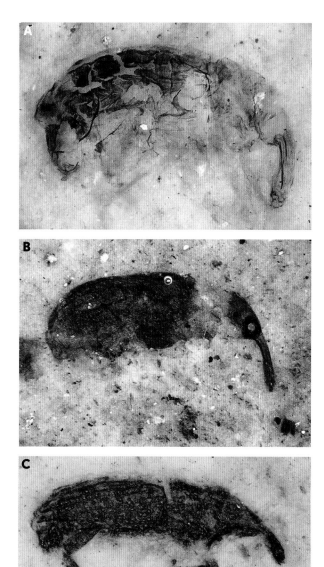

WEEVILS

Easily recognised by their elongated snouts, weevils (superfamily Curculionoidea) are the most common beetles at Foulden. Unlike other types of beetle fossils, weevils have typically been found preserved on their sides (laterally compressed). This is because their stout long-legged bodies tipped over on their side as they sank through the water to the lake floor.

Weevils are also the most abundant beetles at the nearby, slightly younger Hindon Maar *Fossil-Lagerstätte*, where spined weevils similar to the extant New Zealand genus *Scolopterus* are common.[17] These fossils indicate that weevils were abundant and likely ecologically important components in early to mid-Miocene forests, just as they are today in New Zealand forest ecosystems.

The only weevil fossil formally described from New Zealand is the extinct species *Perroudia manuherikia* from an early Miocene (19–16 Ma) fossil site near Bannockburn in Central Otago.[18]

A|B|C Three undescribed fossil weevils from Foulden Maar, all with lateral preservation (lying on their sides). **D** Extant weevil from Herekino Forest, Northland, for comparison. UK

ROVE BEETLES

A new extinct species of rove beetle is one of the exciting finds at Foulden.

Rove beetles, the most numerous beetle family (c. 63,000 species), are found in almost every terrestrial habitat around the world. Their relatively short wing cases cover only part of the abdomen, and most species are small (under 10mm), with a long or sometimes oval body.

They are also the most diverse beetles in New Zealand, with 1230 mostly endemic species from 17 subfamilies known. Most live as predators in decayed wood, plant debris and leaf litter on the forest floor. Many of the major groups of rove beetles found elsewhere in the world are absent from New Zealand, whereas several other groups are diverse here.

Fossils of three genera of rove beetle have been found at Foulden, all representing ancient lineages that are likely to have been part of the forest-floor biota since Gondwanan times: *Sagola* (Pselaphinae), which today has some 130 endemic species in New Zealand; *Nototorchus* (Osoriinae), with two extant New Zealand species; and *Sphingoquedius* (Staphilininae), also present with at least two extant species.

One of these fossils has recently been described formally as a new extinct species. *Sphingoquedius meto* belongs to the tribe Amblyopinini, which is the dominant group among living temperate rove beetles in the Southern Hemisphere.[19] The fossil is 5mm long, medium-sized for its group. The species name is from the Māori verb meto, meaning 'to become extinct'. It has larger eyes than similar living species, a distinct microsculpture on the pronotum, and patches of radiating hairs (setae) on the rear body segments. It would have been a small predator of tiny invertebrates in forest-floor litter.

Only three other fossils of the rove beetle tribe Amblyopinini are known globally – two from amber in the Dominican Republic and one doubtful species from Colorado.

A|B A photomicrograph and interpretive line drawing of *Sphingoquedius meto*, the first fossil rove beetle described from New Zealand and the first fossil of the tribe Amblyopinini from the Southern Hemisphere. **C** Extant amblyopinine rove beetle *S. strandi* for comparison. UK, Josh Jenkins Shaw (Natural History Museum of Denmark)

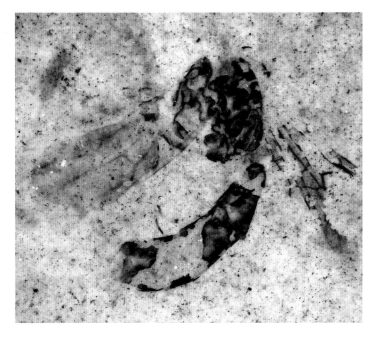

A partially preserved, as yet undescribed Hymenoptera fossil, probably a wasp. UK

ANTS AND WASPS – ORDER HYMENOPTERA

The discovery of several extinct species of ants at Foulden Maar has answered a question that has long puzzled entomologists: has New Zealand's ant fauna always been so strikingly meagre? New Zealand has only 48, mostly introduced species in 30 genera, and only 12 species are native.[20] In contrast, Australia has more than 1600 described species in 110 genera, 87 percent of which are endemic.

Several hypotheses have been suggested to account for this low diversity. Perhaps New Zealand has always been too remote and/or isolated for ants to get here; or perhaps they cannot cope with the cooler climate. One explanation suggested that they were more diverse in the geological past, but were reduced by the decrease in land area following marine transgressions and/or the cooling climates of the Pleistocene.

Fossil evidence was needed to decide which of these hypotheses was most likely correct – and our limited excavations at Foulden have already provided an answer: ants were much more diverse in Zealandia in the geological past.

Hymenoptera are a highly diverse insect order containing sawflies, ants, wasps and bees. There are about 150,000 described extant species worldwide, but little is known about some groups, such as small parasitoid wasps. The estimated total number of species is much higher, possibly even exceeding that of beetles (380,000 species).[21]

The New Zealand hymenopteran fauna is poorly known: the estimated number of undescribed species (775–892) exceeds the number of described species (nearly 750). Many groups that are diverse elsewhere are absent here, or are represented by introduced species only.[22]

About 50 Hymenoptera fossils have been found in New Zealand, including 39 from Foulden, where they are the third most common insects. The wasps appear to belong to the family Ichneumonidae and the superfamily Chalcidoidea, but unfortunately most are too incomplete or too poorly preserved to be identified.

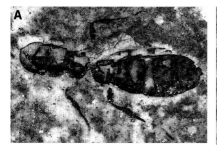

A A reproductive female (gyne) of an indeterminable genus in the subfamily Amblyoponinae.
UK and the late Gennady Dlussky (formerly Moscow State University)

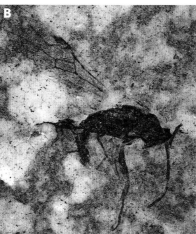

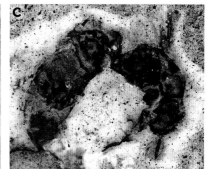

Gynes (queens) of *Rhytidoponera waipiata* (**B**) and *R. gibsoni* (**C**), two extinct ant species in the subfamily Ectatomminae.
UK, GD

D A rare worker ant from the Foulden diatomite, *Austroponera schneideri*, subfamily Ponerinae. UK, GD

Ants are truly social insects, working together to collect and share food and to build and protect their colonies. A typical ant colony comprises one or more winged fertile females (queens or gynes), tens to thousands of males (drones) and similar numbers of sterile wingless workers and soldiers. Winged reproductives, both male and female, are produced over a short period and fly from the nest in a brief mating flight. Fertilised winged females (alates) then drop to the ground, bite their wings off at the base and look for a nesting site.

We now know that by 23 million years ago, in this one small site (a few hectares of rainforest), eight species of ants in five genera and five different subfamilies were established in New Zealand. The true ant diversity at Foulden Maar was actually higher, as only one third of all specimens have so far been described formally, and some of those that remain definitely represent different species.

As with termites, most of the ants found are reproductives or their isolated wings, which fell or were blown into the lake during nuptial flight. The ants described so far include a winged male and one gyne of Amblyoponinae, two gynes in the genus *Rhytidoponera* (subfamily Ectatomminae), a worker in the genus *Austroponera* (subfamily Ponerinae), and a male in the genus *Myrmecorhynchus* (subfamily Formicinae).[23] They all belong to extinct species known only from Foulden.

These are the first fossil records in the world for *Austroponera* and *Myrmecorhynchus*, and the first record of *Myrmecorhynchus* beyond Australia, showing that it once had a wider distribution than today. Three other isolated forewings are from different species, one probably in the subfamily Dolichoderinae.

These fossils represent the first evidence of ants in New Zealand, and they offer a preliminary account of ant diversity 23 million years ago. Earlier assumptions that ants failed to recolonise New Zealand after the Oligocene marine transgression can now be discounted. The taxonomic differences between the Foulden ant fauna and the 12 native species that inhabit New Zealand today suggest that climate cooling since the late Miocene most likely caused loss of diversity.

Two further ant fossils have recently been discovered in New Zealand, one in Miocene amber from Roxburgh in Central Otago (subfamily Dolichoderinae), and one unidentified winged reproductive at Hindon Maar. Both are yet to be described in detail.

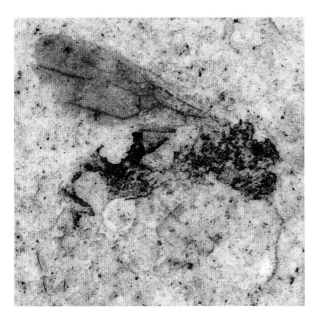

The extinct ant *Myrmecorhynchus novaseelandiae*, a male, in the subfamily Formicinae. UK

CADDISFLIES – ORDER TRICHOPTERA

Trichoptera (meaning 'hairy wings') is a diverse order of primarily aquatic insects that are closely related to butterflies and moths (Lepidoptera) but have wings covered with fine hair instead of scales. The small to medium-sized adults are generally short-lived flying insects, moth-like in appearance and mostly nocturnal. They have mouthparts adapted for the ingestion of liquid foods such as nectar and honeydew. Many species do not feed as adults, however, and they die soon after mating and laying eggs.

The long-lived larvae are aquatic and eat small arthropods and decomposing plant matter and algae. In many species, the larvae construct protective cases or stationary shelters from silk, gravel, sand, plant detritus or other fragments.[24]

Of the 618 extant genera and 16,300 species of caddisflies known worldwide, 29 genera and 244 species (plus some undescribed species) live in New Zealand, representing 16 of the world's 51 families.[25] Most New Zealand caddisflies are endemic and are ecologically important components in streams, trickles and seepages, from coastal areas to the alpine zone. Only a few species inhabit lakes, and five are marine, living in intertidal or littoral habitats.

In the Foulden diatomite, caddisflies are represented only by seven larval cases, some with the larvae in place and some empty. The cases are tubular, straight or slightly curved, 8–23mm in length, and made either entirely of plant debris or a mixture of plant debris and sand-sized mineral and rock grains. The cases alone are of little diagnostic value, but where the larvae are intact they are well preserved and generally clearly show the head, antennae, legs and pronotum, with the rest of the body hidden inside the case.

Several of these fossil larvae, with their elongated heads and long banded legs, belong to the family Leptoceridae (longhorned caddis), some most likely in the extant genus *Triplectides* (stick caddis). This worldwide family has 13 endemic species in New Zealand and, unlike most other caddis, can be dominant in lakes.

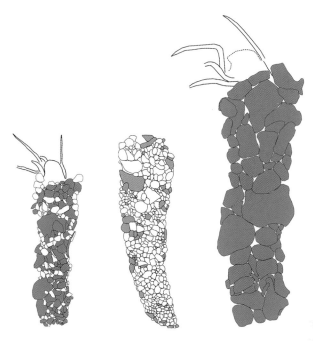

Drawings of caddisfly larvae and their cases from the Foulden diatomite. The larval cases are made from plant debris (grey) and/or sand (white). UK

The larva of a longhorn caddis (Leptoceridae, probably *Triplectides*) preserved in a protective case built of plant debris. Note the banded legs. UK

A rare adult fly from the Foulden diatomite, with overlapping wings, partial legs and part of its body preserved. UK

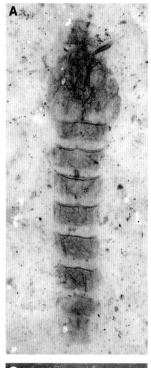

A|B Pupae of non-biting midges (Chironomidae) are the most abundant aquatic insects in the Foulden diatomite.
C Pupa of the extant species *Chironomus plumosus* from Greece for comparison. UK, Viktor Baranov (University of Munich)

Another family probably present at Foulden is Oeconesidae, which is known only from New Zealand (23 species) and Tasmania (one species), where their larvae live in bogs, seepages and smaller streams. Because the Foulden lake was most likely disconnected from streams, the caddis larvae must have lived in the shallow waters of the lakeshore or in small seepages.

No examples of adult specimens, which must have lived nearby, have yet been found in the diatomite.

FLIES – ORDER DIPTERA

Diptera are the second most common insect order after beetles (Coleoptera) at Foulden Maar, and account for 25 percent of all identified insects. This abundance is no surprise, given the extreme diversity of Diptera globally and in New Zealand ecosystems. However, it is likely that the semi-aquatic life cycle of flies led to their over-representation in the diatomite.

Diptera (meaning two-winged) are flying insects in which the hindwings are modified to small club-shaped structures (halteres) that bear sensory organs. Commonly known as 'true flies', Diptera have more than 150,000 described species worldwide, including houseflies, sandflies, mosquitoes and midges.[26] The New Zealand fauna includes 621 genera and about 2500 described species (plus another 750 or so known undescribed genera and up to 1600 undescribed species). About one third of the recorded genera and 91 percent of species occur only in New Zealand.[27]

The majority of dipterans at Foulden are immature aquatic pupae that lived in their thousands in the maar lake and were thus more prone to drowning than fully terrestrial insects. Most of these fragile and minute (5–7mm-long) pupae preserve delicate anatomical details that allow identification to at least family level. A few adult flies found at Foulden include complete individuals and isolated wings that either fell or were blown into the lake from the surrounding rainforest. Prior to research at Foulden, the only Diptera fossil known from New Zealand was the marchfly larva *Dilophus nigrostigma* from early to mid-Eocene (56–38 Ma) siltstones near Livingstone in North Otago.[28]

NON-BITING MIDGES

Chironomidae are a large worldwide family of mostly small delicate flies known as non-biting midges. Adults typically live near bodies of water, mostly freshwater but also saltwater in intertidal zones, and often form large mating swarms at twilight. With a few exceptions, larvae and pupae are aquatic and live either in debris at the bottom or on vegetation; they can be an important food source for fish and other aquatic insects.

About half of the 67 Chironomidae genera occurring in New Zealand are shared with other landmasses, but 94 percent of species are endemic to the main islands or offshore islands of New Zealand, where they occupy a wide range of freshwater habitats.[29]

At Foulden Maar, thousands of immature chironomids may have flourished in the well-oxygenated upper water column of the lake, but some unfortunate individuals sank to the anoxic lake floor to become fossilised. Although their pupae are the most abundant dipterans in the diatomite, no examples of terrestrial adults have been found to date. Species from at least three subfamilies are represented – Chironominae, Orthocladinae and Podonominae. This indicates that non-biting midges were a diverse and perhaps ecologically important component of the lake ecosystem.

PHANTOM MIDGES

Among the insects from Foulden are several pupae of phantom midges (Chaoboridae), sometimes also called glassworms due to their nearly transparent larvae. Chaoboridae are a small family of fairly common midges with a cosmopolitan distribution, but they are absent in today's New Zealand fauna. However, our findings at Foulden Maar have revealed that phantom midges were present in early Miocene freshwater ecosystems in New Zealand, and that representatives of this family subsequently went extinct here.

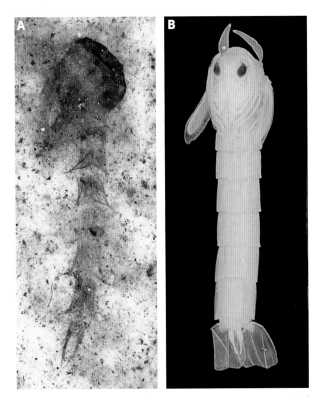

A Fossil of an aquatic pupa of a phantom midge (Chaoboridae). **B** Pupa of an extant phantom midge from Germany for comparison. UK, Viktor Baranov (University of Munich)

C Most macroscopic plant and animal fossils at Foulden are discovered by splitting diatomite on site. In order to study the microscopic organic fraction at Foulden, sediment samples were sent to Nick Butterfield at the University of Cambridge, who used hydrofluoric acid to dissolve the silica component of the diatomite. The organic residue consisted of amorphous organic matter, fungi on leaf cuticle and microscopic remnants of plants and arthropods, including this larva of a phantom midge (Chaoboridae) and other midge larvae. Nick Butterfield (University of Cambridge)

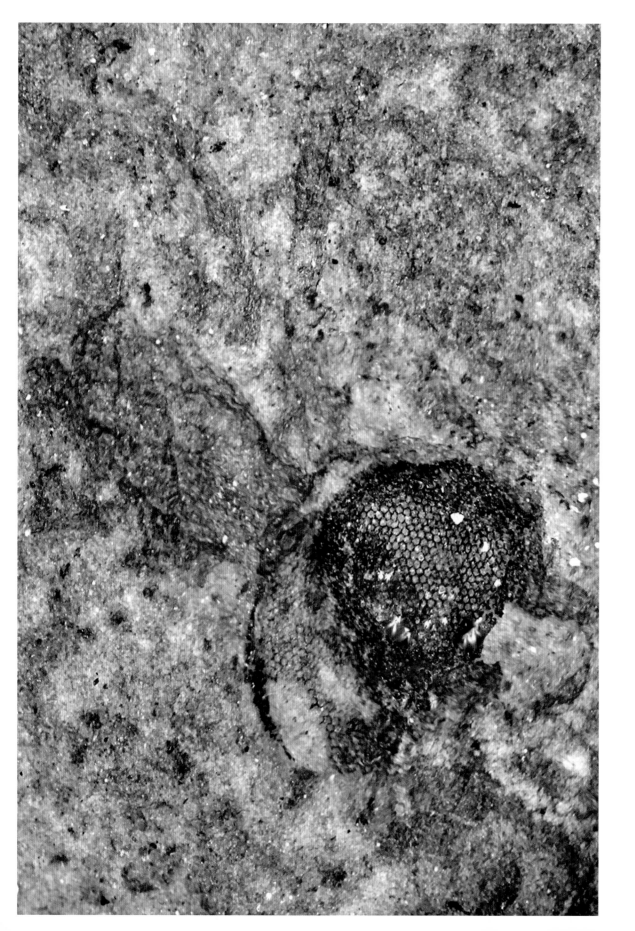

CRANEFLIES

Craneflies are a group of slender long-legged flies of variable size, from a few millimetres up to 5cm, with worldwide distribution. Also known as daddy longlegs, they are ranked as either superfamily Tipuloidea or family Tipulidae, and their phylogenetic position within Diptera is the subject of ongoing research. The oldest known cranefly fossils are from the Triassic, about 240 Ma, and more than 15,000 living species are known globally.[30]

New Zealand has a diverse, highly endemic cranefly fauna that has a relatively high proportion of flightless species.[31] Most species prefer damp or wet temperate environments and are found in forest vegetation near streams or lakes.

At Foulden, craneflies are represented by adult individuals or their isolated wings; no larvae or pupae have yet been found. Although these specimens exhibit typical cranefly characteristics, their incomplete state and poor preservation make closer identification difficult.

DAGGER FLIES

The family Empididae is represented at Foulden by a nearly complete adult male (missing its head and appendages) that is yet to be formally described.

Dagger flies are diverse in New Zealand and include 287 species (all endemic) in 41 genera (plus over 100 undescribed species). They are common in vegetation or open habitats near water, including beaches.[32] Adults and larvae are usually predators of small insects, mostly other flies; adults in some species may also feed on nectar. Some living species form dancing swarms over water or take prey from the water surface – including perhaps the male adult specimen that drowned in Foulden Maar.

Although fossil dagger flies are not uncommon, particularly in Cenozoic ambers, the Foulden specimen is only the third fossil record from the Southern Hemisphere; both previous records are from the Cretaceous (~90 Ma) in Botswana.[33]

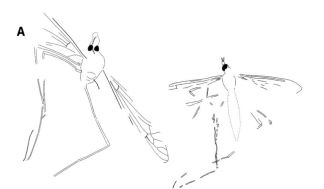

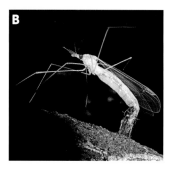

A Line drawings of two adult tipuloid flies, probably craneflies (Tipulidae), from Foulden. Although incompletely preserved, these flies retain some fine details such as the compound eyes (**OPPOSITE**).
B An extant adult cranefly emerging from its pupa, for comparison. UK

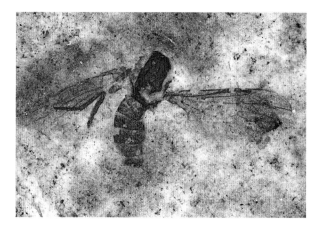

A fossil male adult dagger fly (Empididae), found at Foulden. UK

ARTHROPOD–PLANT INTERACTIONS

Arthropods and plants interact in many different ways. Flowering plants are pollinated by insects; herbivore arthropods feed on various parts of plants; plants provide habitats and may produce structures as shelter for insects; and carnivorous plants have developed mechanisms for trapping arthropods as nutrient sources. Direct evidence for such interactions in the geological past is rare. At Foulden it is represented by the female scale insects found attached to a leaf.

Although there is no doubt that insects played a key role in the pollination of many plant species now fossilised at Foulden Maar, we do not know which insects pollinated which plants.[34] The identification of pollen grains attached to fossil bees at two Eocene maar lake deposits in Germany provided extremely rare direct links, but nothing similar has yet been found at Foulden.[35]

However, the fossil flora exhibits plenty of indirect evidence of arthropod–plant interactions, such as various types of plant damage inflicted by insects; and domatia – structures that plants produce for

A

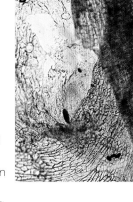

B

A|B The black dots on these fossil leaves show the tiny cavities (domatia) produced by plants to protect beneficial arthropods such as mites. The insect feeding holes seen on a *Hedycarya* leaf (**C**) are surrounded by dark rims of scar tissue. Extensive margin feeding and hole feeding by insects have removed large parts of another leaf **D**. JMB, UK

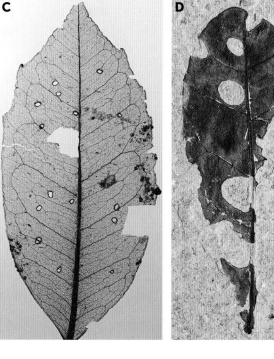

C

D

arthropods (typically mites or ants) with which they live in a symbiotic relationship. These domatia are often small cavities, partly enclosed by hairs or leaf tissue, on the undersides of leaves at the juncture of the midrib and veins. Arthropods inhabit domatia for shelter and, in turn, protect the plant from leaf-devouring fungi; their wastes also provide the plant with nutrients. Two types – hairy tuft and pocket domatia – have been found on fossil leaves at Foulden, although no associated insects have yet been found in situ.

About a quarter of fossil leaves and some seeds from Foulden display insect damage. The most common types are related to external foliage feeding, including variously shaped and sized incisions and perforations at the leaf margin (margin feeding) or within the leaf (hole feeding), typically surrounded by distinct reaction rims of thickened scar tissue.[36] In some leaves the tissue between veins has been chewed up completely (skeletonisation), or patches of the surface tissue have been abraded or removed (surface feeding).

Leaf galls (growths) formed by herbivorous insects during feeding or egg-laying are also present, as are distinct meandering tunnels between the epidermal layers (leaf mines) caused by insect larvae. Other signs of damage include 2–3mm-wide circular or oval punctures caused by piercing and sucking insects.

Tiny circular borings on Menispermaceae (moonseeds) and other seeds from Foulden are most likely the exit holes of insect larvae that were growing in and feeding on the seeds' internal tissue. Although it is impossible to assign the various damage types to particular insect culprits, their overall variety indicates a much higher diversity of phytophagous or plant-feeding insects at Foulden than is known from the insect fossils found thus far.

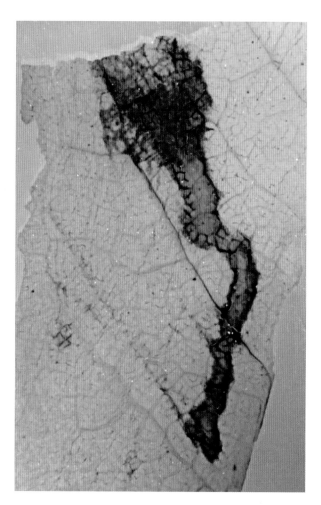

Example of a leaf mine on a fossil leaf from Foulden. Moths, sawflies and true flies are the most common leaf miners. Their larvae feed on the leaf tissue, leaving faeces or frass, and widening the mine as they grow. JMB

WHY THE FOULDEN CLIMATE RECORD IS SO EXCEPTIONAL

- The maar was a small, deep, isolated lake on a low-lying landmass surrounded by the Southern Ocean, and accumulated finely laminated organic sediments with little influx of extraneous material.

- Tectonic stability in the region and the absence of burrowing animals or water movement on the anoxic lakebed meant the sediment archive lay undisturbed for 23 million years.

- The maar was situated at 47°S (it is now at 45°S), about halfway between the equator and the South Pole, a latitude that experiences climatic influences from both tropical and polar regions.

- The maar's continuous climate record relates to the Oligocene–Miocene boundary, when major changes in the Antarctic ice sheet took place over a relatively short time.

- The climate record shows annual seasonal changes in minute detail.

- The maar formation is precisely dated at just before 23 Ma, from sources including radiometric dates of fine-grained volcanic rocks associated with the eruption, palynology, and the serendipitous discovery of evidence for a reversal of the Earth's magnetic field near the base of the core.[1] In most deposits, scientists are hard pressed to obtain even one of these age determinants.

- The exquisite preservation of leaf fossils, including microscopic anatomy, can be used to reconstruct paleotemperatures, rainfall, seasonality and changing CO_2 levels in the early Miocene atmosphere.

- The preservation of organic molecules can be used to determine lake chemistry and hydrology.

CHAPTER 7

CLIMATE SIGNALS FROM THE DISTANT PAST

Possibly the greatest concern of the twenty-first century is a warming and unstable climate. Over the past 200 years humans have been releasing huge amounts of CO_2 and other gases such as methane, which trap heat in the Earth's atmosphere. CO_2 levels have risen steadily from ~300ppm at the beginning of the Industrial Revolution to ~410ppm in 2021, and humankind is struggling to grapple with the likely environmental effects.

Scientists are developing models to predict climate trends. One important aspect of this involves looking back into the distant past, to a time when the Earth was warmer and CO_2 levels were higher than they are today. Only a few places on Earth can provide the detailed records required for this study.

The most continuous climate records come from cores drilled through the thick ice sheet in East Antarctica. These have provided high-resolution (year-by-year) data on temperatures and corresponding CO_2 levels. However, the ice record goes back less than a million years, and CO_2 levels were consistently below 300ppm during this period.

Deep-ocean sediment cores can give a longer perspective – over tens of millions of years. But they provide little detail, as a 100m-long core may span a million years or more and may have gaps caused by periods of erosion or no sediment deposition.

In a very few select places on Earth, cores taken from deep isolated maar lake deposits provide precise and continuous details of changing climate for up to a million years at particular periods in geological time. One such site in the Northern Hemisphere is Messel Pit in Germany, where a core sample records climate details for about 640,000 years in the warm mid-Eocene, about 47 Ma.[2]

Foulden Maar is another key site for reconstructing past climate detail, and fortuitously, this site captures events at the Oligocene–Miocene boundary (23 Ma), a time of considerable relevance for predicting future climate changes.

EARTH'S CURRENT STATUS: INTERGLACIAL

Earth is currently in an interglacial period that began after the Last Glacial Maximum ended, about 14,000 years ago. Antarctic core records of the past 800,000 years show us that there has been a series of warm peaks and cool troughs matching the numerous glacial and interglacial cycles that Earth has been experiencing since at least the start of the Pleistocene, 2.6 million years ago.[3]

FOULDEN MAAR AS A KEY CLIMATE INDICATOR

Foulden Maar is the only site of pre-Quaternary age in the Southern Hemisphere with an annually resolved climate record, something few – if any – potential sites in the world are capable of providing.

At Foulden, the sedimentation rate was around 1000mm per 1000 years.[4] This provides a chronology (time record) at a resolution more than 10 times higher than that preserved in deep-sea cores, which are the other main source of paleoclimate records for the earliest Miocene.

The exquisite preservation of leaf fossils at Foulden, including microscopic anatomy, can be used to reconstruct paleotemperatures, rainfall, seasonality and changing CO_2 levels in the atmosphere. Organic molecules can be used to determine lake chemistry and hydrology.

ESTIMATING TEMPERATURE AND RAINFALL IN THE EARLIEST MIOCENE

Several different methods have been used to estimate paleotemperature from the plant fossils recovered from Foulden Maar. The first is a quantitative comparative method known as the Climate Leaf Analysis Multivariate Program (CLAMP). Devised in the 1970s and constantly updated, CLAMP uses correlations observed between climate variables and multiple leaf characteristics of modern flowering-plant leaf assemblages from hundreds of reference sites all over the world. The association between the frequency of certain leaf features and different components of the climate can be used to infer climatic conditions.

One of the main advantages of this method is that it is not necessary to identify the leaf taxonomically. Instead, the approach is based on characters such as leaf size, ratio of length to width, shape and lobing, the form of the base and type of apex, whether the leaf has a smooth (entire) margin or is toothed, and the regularity of the teeth as well as their closeness, shape and spacing.[5] These characters are measured for 20 or more leaf types from one fossil site. Each leaf is scored, and the values can then provide a modern location that has a similar climate to that fossil site.[6]

The second approach, known as bioclimatic analysis, depends on identifying the families and genera of the leaves. The method then uses the climatic envelopes of the nearest living relatives for all identified taxa in the fossil flora to determine the climatic range in which the majority of those plants could co-occur. The high preservation quality and diversity of leaves at Foulden Maar have allowed this method to be applied at the site.

A third approach depends on detailed measurement of the features of certain distinctive conifer leaves. Australian researchers had earlier reported a correlation between the leaf surface area of broad-leaved *Podocarpus* and the mean annual temperature of a sample site.[7] Because large leaves of *Podocarpus travisiae* are quite common at Foulden, measurement of their surface area became an additional way to estimate past temperatures.

The application of all three methods has revealed that the climate at Foulden in the earliest Miocene was humid, mesothermal (warm temperate to marginally subtropical), with some rainfall seasonality.[8] Mean annual temperature was about 18°C (similar to modern-day Brisbane) without major extremes. The temperature for the three warmest months averaged 23°C and the three coldest about 13°C. To put these estimates into context, the current mean annual temperature for Middlemarch (near Foulden) is about 10°C, with an average for the warmest months of 17°C and for the coldest of 8°C. Thus, we know that 23 million years ago the mean annual temperature at Foulden Maar was about 8°C warmer than it is today. It was most likely frost-free with no snow. Returning to the climate of 23 million years ago may seem a pleasant prospect for Middlemarch and southern New Zealand as climate warming continues on Earth today, but the outcome of that level of warming in other areas of the planet would be disastrous.

Rainfall levels, which are also derived from the CLAMP methodology, were high enough to sustain rainforest (more than 1500mm per year and possibly 1700–2000mm per year – wetter than modern-day Dunedin). The potential for summer moisture deficits was also determined (supported by the paucity of epiphyllous fungi at the site). In comparison, rainfall around Middlemarch today is about 400mm per year, with rain days spread throughout the year. Analysis of the sediments in the Foulden Maar core suggests that over the course of the build-up of the lake sediment, rainfall in southern New Zealand was variable, subject to the westerly wind belt and tropical ocean currents from the north.[9]

FOULDEN FINDINGS CHANGE THE PICTURE

Foulden Maar formed after the first Antarctic glaciations (35 Ma) but before the ice ages of the Quaternary (2.6 Ma), and just after the maximum marine transgression of Zealandia, when the land area in New Zealand was at its smallest. Since we know Foulden Maar was at approximately the same latitude as it is today, what was responsible for this markedly warmer climate?

Warmer temperatures in the distant past are related to higher levels of CO_2 in the atmosphere. We now know, for example, that during the early Eocene

BELOW Global average surface temperatures through the Cenozoic, calculated using information obtained from deep-ocean cores. The graph shows many fluctuations, with an overall decrease from highs in the Eocene and a drop in the early Oligocene corresponding to the beginning of stable ice sheets in Antarctica. Another high in the mid-Miocene was then followed by a generally downward trend to the present day.[10] Adapted by Tammo Reichgelt

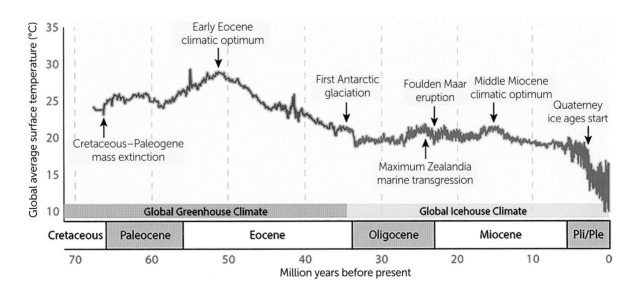

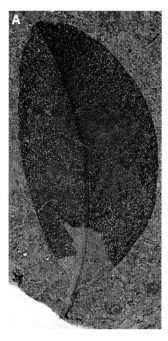

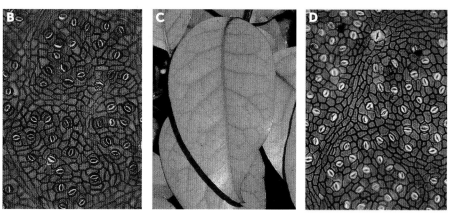

Stomatal measurements from *Litsea calicarioides* leaves (**A**|**B**), one of the most abundant species at Foulden, were compared with those from the closely related living New Zealand species *L. calicaris* (mangeao) (**C**|**D**). The results showed that Foulden leaf-based CO_2 levels of 450–550ppm are considerably higher than previous marine-based estimates. This helps to provide an explanation for the globally higher temperatures of the early Miocene.[16] Tammo Reichgelt, JGC

(56 Ma), the warmest time in the past 70 million years, atmospheric CO_2 was at least 1000ppm.[11]

Previous estimates for the early Miocene have indicated modest atmospheric CO_2 levels of ~300ppm, similar to pre-industrial values.[12] This 'early Miocene CO_2 paradox' has been difficult to reconcile with the markedly higher temperatures globally. However, most previous estimates were based on marine proxies rather than on-land evidence such as at Foulden Maar.

Leaves, which are of course in immediate contact with atmospheric CO_2, respond directly to concentration changes because they use the gas for photosynthesis, producing starch and other energy-rich carbon-based materials. CO_2 enters the leaves through stomata (pores), usually found on the underside of the leaf. In many species the density of these pores declines as CO_2 levels increase, because the plants do not need as many pores to get the necessary amount of CO_2.

The relationship between leaf pore density and CO_2 levels has been observed directly in greenhouse experiments; it can be seen in fossil leaves. Importantly, the rapid mixing of CO_2 in the

atmosphere globally means that values measured at one site can be extrapolated to the rest of the planet.

The mummified leaves from Foulden Maar, from right throughout the core, are so well preserved that minute details of venation, cuticles and stomata can be measured. As a result, this site presents a very rare opportunity to make measurements of leaf anatomy that translate directly to climatic and atmospheric conditions at the time those leaves were growing.[13] This allows scientists to track changes in atmospheric CO_2 through time, from the bottom of the core (450ppm) to the middle (550ppm) and the top (back to 450ppm).[14] These changes may be related to the reduction in the ice sheets in Antarctica, which went from 125 percent of present-day volume to about 50 percent of the modern extent at the Oligocene–Miocene boundary.[15]

Another important finding from the leaves in the Foulden core is that the trees in the rainforest around the maar behaved quite differently from trees today. Leaves need stomata to take up CO_2 from the atmosphere, but they also lose water to the atmosphere through these stomata. Anatomical and chemical research indicates that the 'fossil' trees were better at conserving water than modern trees living in much more water-stressed environments.

This suggests that plant growth may improve and forests may expand as CO_2 levels increase – a phenomenon known as 'global greening'. In general, a 'greener' world sounds positive, in particular because of the values we associate with the word 'green'. However, it is important to remember that the change in water-use efficiency took millions of years and resulted in globally changed ecosystems, with many new species that thrived within them. Going back to the 'greener' world of 23 million years ago could result in a similar reshuffling of life around the world and the extinction of many life forms – but over the course of decades and centuries rather than millions of years.[17]

EL NIÑO AND LA NIÑA

El Niño and La Niña are both phases of the El Niño Southern Oscillation (ENSO), a naturally occurring global climate phenomenon that affects rainfall, wind, temperature and even the strength and frequency of tropical storms. In recent years, a great deal of attention has been focused on understanding ENSO dynamics in order to understand how it will respond to increased climatic warmth.

During an El Niño phase nowadays, New Zealand experiences stronger westerly winds in summer that bring higher rainfall to western areas, drier weather to the east and colder winter temperatures. In a La Niña phase, ocean temperatures in the east and central tropical Pacific drop, because stronger trade winds bring cool deep water to the ocean surface. When this happens, sea temperatures can warm above average in the far western Pacific, including around New Zealand.

These cycles are not new. ENSO-like cycles (of two to eight years) have been detected in many paleoclimate records for the Holocene – the past 10,000 years – but we do not have reliable records going back further. Using the detailed sediment record at Foulden, however, we were able to detect a very strong ENSO signal – an indication that it was a major influence on the New Zealand and Southern Hemisphere climate 23 million years ago.[18] Whether the climatic effects of El Niño and La Niña phases mirrored those of today is one of several outstanding mysteries of Foulden Maar that remain to be solved.

WHAT DOES THIS TELL US?

Foulden Maar has allowed climate scientists to break new ground in studying the paleoclimate. The core, which was taken through the thickest part of the diatomite deposit, provides a more than 120,000-year-long record of the annual climate conditions in its seasonal sediment layers. This is the only known uninterrupted and undisturbed annual and even seasonal record for the earliest Miocene period on Earth, making it possible to study individual years and seasons, and to perceive changes on a human time scale.

So far the archive shows that the mean annual temperature at Foulden was about 8°C higher than in southern New Zealand today, and rainfall was greater. This high precipitation and increased warmth supported rainforest, in contrast to the cool open dryland vegetation that occupies the area today (or did so until it was modified after the arrival of humans about 800 years ago).

Foulden Maar is also the only site of Miocene age in the Southern Hemisphere that preserves an ENSO record, demonstrating that ENSO was a major, if not the dominant source of climate variability in earliest Miocene, as it is in New Zealand today.

Studies of atmospheric CO_2 levels from the mummified leaves in the core show that they ranged from 450 to 550ppm in the earliest Miocene, much higher than previous estimates of about 300ppm from studies elsewhere in the world.[19] These values are similar to the CO_2 levels predicted to be reached on Earth by 2040–50, and demonstrate the dangerous increase in global temperatures that is likely to accompany the projected increase in CO_2.

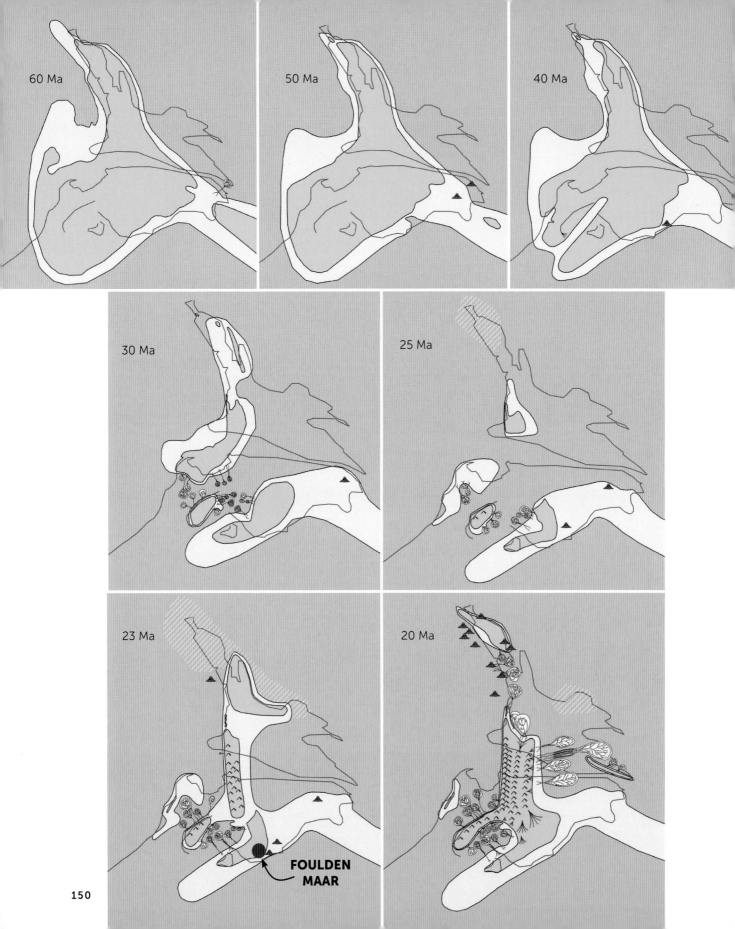

60 Ma

50 Ma

40 Ma

30 Ma

25 Ma

23 Ma

FOULDEN MAAR

20 Ma

CHAPTER 8

A UNIQUE FOSSIL SITE AND ITS FUTURE

Until the late Cretaceous, about 83 Ma, the continent of Zealandia, including present-day New Zealand, was a large landmass (4.9 million square kilometres, about the size of modern-day India) on the eastern margins of Gondwana, most of which was above sea level.[1] It shared a fauna and flora with Australia and Antarctica that included dinosaurs, tuatara, frogs and conifer/*Nothofagus* forests on land, and mosasaurs, plesiosaurs and ammonites in the oceans.[2] Some of these animals – including ammonites, all the dinosaurs and large marine reptiles such as plesiosaurs and mosasaurs – became extinct at the end of the Cretaceous (see time scale on p. 13).

After the gradual opening up of the Tasman Sea, which began in the south about 83 Ma, Zealandia drifted east and north. It reached its present latitudinal position in the Eocene, by which time any land connections with Australia had been lost.[3]

During the late Oligocene the land area was reduced by inundation to a series of large low-lying islands occupying an area of about 20,000 square kilometres (comparable to present-day New Caledonia), which supported a diverse biota.[4] Around 25 Ma, in the late Oligocene, a new plate boundary between the Australian and Pacific plates became active and initiated uplift, gradually increasing the land area to nearly its present size of 268,021 square kilometres.[5]

Situated near the southeastern coast of Zealandia, Foulden Maar erupted in an area where terrestrial volcanic activity began around 24 Ma and continued intermittently for 10 million years. Foulden Maar provides an astonishingly detailed window into lake and forest ecosystems at this critical time in Zealandia's history, contemporaneous with, or soon after maximum marine transgression.

THE 'OLIGOCENE DROWNING'

In recent decades, debates about the antiquity of the plants and animals native to New Zealand have focused on the so-called 'Oligocene drowning' (33–23 Ma), a time when much of the land now known as New Zealand was extensively submerged as marine environments expanded.

Traditionally, we have assumed that terrestrial environments always existed over the last 100 million years. This long-term continuity is reflected in the presence of many ancient elements that have persisted since the breakup of Gondwana about 83 Ma: lineages such as velvet worms, kauri trees and tuatara were thought

OPPOSITE

Paleogeographic maps showing the changing shape and size of Zealandia from 60 to 20 Ma. Green indicates land, pale blue is shallow sea and darker blue is deep sea. The red dot shows the location of Foulden Maar.

P.J.J. Kamp (University of Waikato), unpublished New Zealand paleogeographic maps used with permission

to be evidence of continuous land.[6] However, these assumptions would be overturned if New Zealand had at any time been completely under water – if maximum marine transgression was in fact *complete* marine transgression, as some have suggested. In that case, our biota would be relatively young and entirely sourced from other landmasses.[7]

For geologists, key questions have been: How much land remained above sea level at the time of maximum marine transgression? and: Is there any evidence in the fossil record of major turnover in plant and animal life at this time? These ideas have generated many research papers that have attempted to explain the absence of mammals and snakes in New Zealand when they are ubiquitous across the Tasman, and the apparent dearth of some insect and other groups.

The terrestrial biota preserved in the Foulden Maar diatomite has the potential to answer these and many related questions. It is dated precisely to 23 Ma, a time when peak 'drowning' of New Zealand is thought to have occurred.[8] The fossil treasures, as well as those from nearby fossil sites both slightly older and slightly younger, confirm that New Zealand supported a diverse range of plants and animals throughout maximum marine transgression. These fossils have strong links to older (Gondwanan) lineages, and in many cases represent the antecedents of the modern New Zealand biota.

Foulden Maar indicates that there is no biological discontinuity evident in the fossil record, and that terrestrial environments in Zealandia were present throughout the period of marine transgression. The land area changed in extent, but at no time was all of Zealandia synchronously submerged. Recent comparisons of evolutionary studies of animals and plants pre- and post-Oligocene support the conclusion that there was no mass extinction during the period of maximum inundation.[9]

THE MAAR LAKE ECOSYSTEM

Foulden Maar was home to a limited array of plants and animals – a result of its size, insularity and the fluctuating water quality of the ecosystem. This is typical of maar lakes. The aquatic microflora was dominated by a single diatom species, *Encyonema jordaniforme*, which arrived soon after the lake formed and dominated the phytoplankton thereafter. This mucilaginous diatom was the key species in the lake's food chain.[10]

Other algae included the clump-forming *Botryococcus*, spores of which made up almost 6 percent of the microflora at times.[11] A third type of algae, a species of golden-brown Chrysophyceae, is known only from tiny siliceous resting spores preserved in the dark-coloured winter layers of diatomite.[12]

Freshwater sponges that lived on the phytoplankton may have been attached to fallen trees or branches around the lake edge, to rocks, or to the stems of water plants such as raupō, rushes, sedges and water milfoil that occupied shallow or swampy areas around the margins of the lake.[13] As in other maar deposits, there were few species of aquatic insects at Foulden: to date those found include only caddisfly larvae, a possible water beetle and various midge larvae.[14]

Galaxiid fish inhabited the oxygenated upper water column of the lake in large numbers. Larval and juvenile fish lived on diatoms, as seen from the coprolites made up exclusively of diatoms and some organic material. Mature *Galaxias effusus* were presumably ambush predators, capturing terrestrial and aquatic arthropods and ingesting some plant detritus, sponges and mineral grains as they foraged in shallow zones at the lake's margin. Top of the food chain was the freshwater eel *Anguilla*, which could have dined on arthropods, smaller fish and possibly birds.[15]

Although no bird fossils have been found, there is ample indirect evidence for waterfowl in the form of coprolites containing quartz sand grains and a variety of different diatoms. As there is no quartz sand in the maar, these grains must have been ingested along with the diatoms as waterbirds foraged in waterways well beyond the maar.

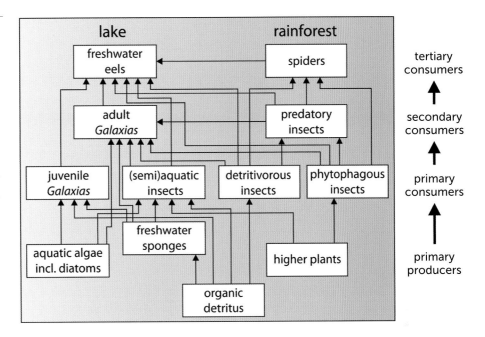

A schematic food web or food chain for Lake Foulden and the surrounding rainforest, showing the most important biological interactions. The arrows show who was eating whom – the pathways of consumption and energy flow in these ecosystems – based on the fossils found to date. UK

Snails with calcareous shells may also have inhabited the lake. However, since acidic conditions in the lake's deepest waters dissolved apatite (a bone mineral) from fish bones, all calcareous material, such as that of potential freshwater molluscs and other calcareous organisms, if it existed, has vanished.

THE RAINFOREST FLORA

Charred pollen grains at the base of the Foulden core record the incineration of the pre-existing rainforest vegetation when the maar-volcano erupted. Tracking up the core, pollen and spores provide an extraordinary record of the initial colonisers of the devastated site (ferns and fern allies), followed by the eventual return of evergreen rainforest to the nutrient-rich volcanic soils around the maar.[16] Lauraceae pollen is absent because it is fragile and does not preserve; *Nothofagus* pollen makes up 7%; that of other eudicots and basal angiosperms just over 50%; fern spores contribute 15%; conifers make up about 13% and monocots 11%.

The leaves tell a different story, however: Lauraceae dominated the flora (44% of all leaves); many different dicots made up about 50%; and conifers and monocots made up the remaining 5%. *Nothofagus* is not represented at all. This data highlights the value of maar fossils, where preservation can capture elements of the resident biota that are missing from other types of fossil sites. Foulden enables us to combine and integrate all available data to get a complete picture of the vegetation at one site in the geological past.

Using a system devised for Australian rainforests, the vegetation that surrounded the Foulden lake is formally classified as a 'simple notophyll vine forest'.[17] The classification is based on a relationship between climate and the average leaf size of the dominant trees. Generally, the warmer and wetter the climate, the larger the leaves and the more complex and diverse the forest.

Part of this complexity is in the development of layering in the physical structure. In warmer rainforests there is a more or less continuous even-height canopy consisting of numerous tree species, with occasional taller emergent trees, an understorey of tree and shrub layers and a ground layer of herbs and tree and

VEGETATION TYPES FOUND AT FOULDEN MAAR

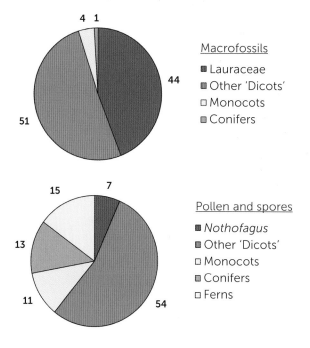

Macrofossils
- ■ Lauraceae
- ■ Other 'Dicots'
- □ Monocots
- ■ Conifers

Pollen and spores
- ■ *Nothofagus*
- ■ Other 'Dicots'
- □ Monocots
- ■ Conifers
- □ Ferns

Diversity of vegetation types (**ABOVE**) and vegetation communities (**RIGHT**) of the Foulden environment 23 million years ago. The top two pie-charts are based on macrofossils only (excluding pollen and spores). The bottom two include pollen and spores only. Note the absence of *Nothofagus* (no macrofossils) and the absence of Lauraceae (no pollen). A relatively high diversity of plants in the different forest or lake-edge habitats or roles is reflected in both the macrofossil and the pollen signals, suggesting that the forest provided a wide range of habitats and niches across a relatively small area. JGC

shrub seedlings. Climbing plants (lianes and vines), mistletoes and tree-dwelling epiphytes (ferns, orchids, etc.) are generally present and often diverse.

The high-light environment of the lake shore also fostered some short-lived, quick-growing tree species such as *Malloranga*; and the shallow lake margins were home to a range of aquatic plants and algae – mostly emergent monocots such as raupō.[18]

We now have detailed information about the composition of the forest around the lake 120,000–130,000 years after the eruption, as most of the plant macrofossils have been collected from the 10m or so of diatomite exposed in the surface mining pits. By this late stage in the forest succession, the canopy trees included two emergent conifers, *Podocarpus* and *Prumnopitys*, along with *Laurelia* and at least 10 species of Lauraceae, which were the canopy dominants.[19] Other trees living close enough to the lake for leaves, flowers and/or fruit to fall in the water included two genera of Cunoniaceae, two genera of Elaeocarpaceae, several types of Euphorbiaceae including the common *Malloranga*, a species of Meliaceae closely related

VEGETATION COMMUNITIES AT FOULDEN MAAR

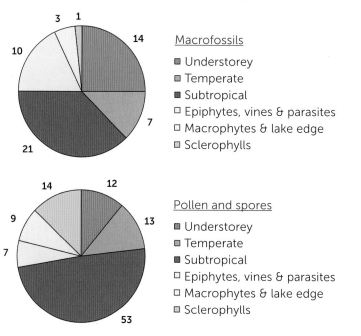

Macrofossils
- ■ Understorey
- ■ Temperate
- ■ Subtropical
- □ Epiphytes, vines & parasites
- □ Macrophytes & lake edge
- ■ Sclerophylls

Pollen and spores
- ■ Understorey
- ■ Temperate
- ■ Subtropical
- □ Epiphytes, vines & parasites
- □ Macrophytes & lake edge
- ■ Sclerophylls

to *Dysoxylum*, a species of *Fuchsia* (Onagraceae), Myrtaceae, at least two different Proteaceae, and species of Rutaceae and Sapindaceae (see Table 1 in Appendix).

The understorey included many shrubs and small trees, vines, lianes, and epiphytes such as ferns and orchids. Smaller trees and shrubs that formed part of the understorey included species of *Alseuosmia*, Araliaceae, Celastraceae, *Cordyline*, *Hedycarya* and Winteraceae. Lianes included several species of *Ripogonum*, Bignoniaceae and Menispermaceae. Ferns, including *Davallia* and a species of Polypodiaceae, were probably epiphytic on trees or rocks on the lake edge; orchids, the perching lily *Astelia* and *Luzuriaga* may have fallen into the lake from branches overhanging the water.

THE RAINFOREST FAUNA

The arthropod fauna in the rainforest was dominated by adults of mainly ground-dwelling taxa of forest-floor, leaf-litter and wood habitats, such as bark bugs, weevils, scale insects, termites and ants. Many fell or were blown or washed into the maar lake, and poor flyers such as the alates of ants may have fallen into the water during swarming.

Three members of the Diptera are the only large-winged insects present. We cannot yet determine whether this reflects the rarity of such taxa in the forest ecosystem, taphonomic bias (the faster decay of some taxa), or incomplete sampling. Since the present diversity estimates are based on fewer than 300 arthropod fossils, we can assume that the actual diversity was at least an order of magnitude greater.[20] Sap-sucking scale insects on Elaeocarpaceae and Lauraceae leaves provide direct evidence of insect–host relationships.[21] Domatia on *Malloranga* and other leaves could have provided homes for tiny mites, and although only one fossil mite has been found to date, Acari (mites) were probably abundant, as they are in modern New Zealand forests.[22] Several genera of termites that shed wings into the lake were common colonisers of dry and wet wood.[23] The diverse ants were

mostly foragers on the forest floor; the rove beetles lived on tiny invertebrates in the forest-floor litter; and bark bugs, as their name suggests, lived under the bark of dead or dying trees.[24]

Damage on many leaves provides evidence of consumption by a wide range of insects. At Foulden this includes leaf margin, hole and surface feeding, galling, leaf mining, piercing and sucking, and seed predation. Many types of insect larvae are implicated in these interactions, and further study might make it possible to link specific insects to particular damage types.[25] Other indirect evidence for insect–plant interaction comes from the number of insect-pollinated flowers present in the flora.[26]

Spiders were the top predators in the forest ecosystem, including a trapdoor spider that would have preyed on passing insects from its burrow.[27]

Overall, the fossil arthropod fauna from Foulden is mainly characterised by taxa with poor flying ability, and thus low potential for dispersal. In combination

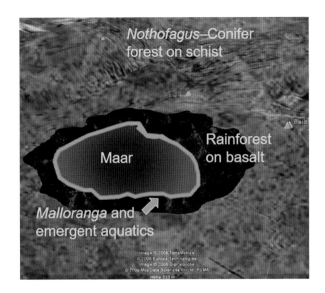

A reconstruction of the vegetation environment and biota at Lake Foulden, with lush warm-temperate rainforest growing immediately around the small lake on fertile volcanic soils, and southern beech/conifer forests on the more distant and poorer surrounding soils. JCG (modified from Google Earth)

with the high arthropod diversity and evidence of arthropod–plant interactions, this indicates a diverse and complex terrestrial paleoecosystem.

Birds must also have been an integral part of the forest ecosystem, as indicated by the abundant bird coprolites in the lake sediments. Many of the fossil fruit from the forest around the lake are quite large and fleshy and were probably eaten by pigeons. Although no birds have been recovered from Foulden to date, fossil pigeons, parrots and many other birds have been found at nearby early to mid-Miocene St Bathans Lake Manuherikia fossil localities.[28] Molecular evidence for the origins of nectar-feeding New Zealand birds indicates a very long history for some endemic groups, which suggests that they could well have been present at Foulden.[29]

BIOGEOGRAPHY AND EXTINCTIONS

The average longevity of animal and plant species is estimated to be between two and 10 million years, so nowhere on Earth are there species that predate the Miocene, although many have close relatives living today. All species that lived in and around Foulden Maar are now extinct, and some of the plant and animal groups at Foulden have also subsequently become extinct in New Zealand. These are mostly biota restricted today to warm wet climates.

About half of the plants at Foulden have close relatives in the modern New Zealand flora. They include the ferns *Davallia* and *Lecanopteris*, the conifers *Podocarpus* and *Prumnopitys*, cabbage trees, supplejack, *Astelia*, *Luzuriaga*, the orchids *Earina* and *Dendrobium*, water plants such as bulrush and water milfoil, and trees such as *Fuchsia*, pigeonwood (*Hedycarya*), *Laurelia* and two of the three genera of Lauraceae (*Beilschmiedia* and *Litsea*). However, even then, the fossil *Podocarpus* is not closely related to the species that live in New Zealand today, but to a large-leaved relative in Australia.

Trees that are no longer part of the New Zealand flora include *Akania*, Celastraceae, *Cryptocarya* (the most common Lauraceae at Foulden), *Fouldenia*, *Malloranga* and two groups of Proteaceae, as well as vines such as Menispermaceae. In addition, numerous unidentified leaves, flowers and pollen hint that many other trees and shrubs were present in the flora. Nearest relatives for most of these taxa live today in Australia, New Caledonia and South America.

The lake inhabitants mostly have close relatives in New Zealand today; and some, such as the *Galaxias* and *Anguilla*, provide the earliest known fossils in the Southern Hemisphere, suggesting a long history in New Zealand. The spiders and insects reveal similarities to, but also notable disparities with, the modern New Zealand arthropod fauna. Trapdoor spiders still live in New Zealand, as do bark bugs, scale insects, rove beetles and many of the insect groups with aquatic larvae. Similarly, the termites found at the site belong to families that are still present in New Zealand. In contrast, most of the ants and some midges belong to groups that are not part of the modern New Zealand biota, and some genera from Foulden are extinct globally.[30]

AFTER THE LAKE DRIED UP

After more than 130,000 years, the maar was completely infilled with sediment. In the final stages it would have been a marshy swamp, before becoming dry ground. This transition can be seen in the maar lakes of different ages and stages in the Eifel region of Germany.

The evergreen rainforests probably continued to occupy the fertile basalt soils of what is now East Otago for the next few million years. We have some insight into the regional vegetation from other, younger maars that formed around 14 Ma at Hindon, 25km to the east. There, preliminary research tells a different story about the biota, although some elements are similar.[31] The diatoms at Hindon are different, and the sediment is a black carbonaceous mudstone with some diatomaceous layers. Hindon has well-preserved *Galaxias* fish and eels, as at Foulden, but a very different assortment of insects. Most importantly, unlike Foulden, the forest

Reconstruction of the early Miocene Lake Foulden. Plants surrounding the lake include the epiphytic fern *Davallia*, a large-leaved *Podocarpus* and the monocots: *Cordyline*, the liane *Ripogonum*, epiphytic *Astelia* and the orchid *Dendrobium* and lake-edge *Typha*. Angiosperm trees include Lauraceae, *Fuchsia*, '*Dysoxylum*', *Laurelia*, *Malloranga*, a species of Myrtaceae and a species of Proteaceae. Arthropods include a spider, ants, bark bugs, beetles, termites, scale insects and a weevil. Paula Peeters

surrounding the Hindon maars was dominated by species of *Nothofagus*. Some Araliaceae, Lauraceae and Myrtaceae are present, and *Ripogonum* is common, but overall the forest ecosystems were very different. Whether this was the result of cooling climate or the varying elevations above sea level of the lakes is still uncertain.

Other volcanic lake deposits closer to Dunedin are younger still, at 12 and 11 million years old. Those at Kaikorai Valley and Double Hill are probably also maars, though this needs to be confirmed by drilling.[32] All of these other deposits are dominated by *Nothofagus*, some podocarps and a variety of flowering plants, and all have preserved galaxiid fish, insects and/ or insect-damaged leaves.[33]

From about 10 Ma the global climate began to cool, and it continued on a cooling trajectory until the onset of the numerous glacial and interglacial cycles of the Quaternary, about 2.6 Ma. This would have sounded the death knell for most of the warmth-loving vegetation that once surrounded Foulden.

By the time the first settlers from Polynesia arrived, almost 1000 years ago, the vegetation around Foulden comprised scrub and open forest of kōwhai, tōtara, matagouri and tussock grasses that grew in a dry, cool temperate climate. Tōtara is almost the only tree growing around Foulden today that has anything in common with the forests of 23 million years earlier; however, the small-leaved common *Podocarpus hallii* is a distant cousin to the large-leaved Miocene *Podocarpus*.

MAARS AND *LAGERSTÄTTEN* DEPOSITS: PUTTING FOULDEN MAAR IN CONTEXT

Scientists employ the term *Fossil-Lagerstätte* to designate rare sedimentary deposits that exhibit exceptional fossil preservation and provide amazing insights into past ecosystems, environments, biodiversity and evolution. There are two types: concentration *Lagerstätten* (*Konzentrat-Lagerstätten*) preserve huge numbers of fossils; and conservation *Lagerstätten* (*Konservat-Lagerstätten*) 'preserve quality rather than quantity'.[34]

Foulden Maar is recognised globally as one of a very few Miocene *Konservat-Lagerstätten*. This reflects the exceptional level of preservation of organic material, including eyes and skin; and the quality of ecological representation, from microscopic algae and aquatic and forest-dwelling insects to entire fish, and the leaves and flowers of forest trees, shrubs and epiphytes, from which it is possible to reconstruct entire ecosystems (see Tables 1 and 2 in Appendix).

Only a few pre-Quaternary maar deposits have been studied in detail. Other than Foulden and the Hindon Maar Complex in New Zealand, most of these are of Cenozoic age and are in the Northern Hemisphere. Several of the best-studied maars are in Germany. Messel Pit (Grube Messel) is by far the most famous maar *Konservat-Lagerstätte*.[35] Two kilometres in diameter and filled with dark laminated oil shale, it has been known as a fossil site for more than 140 years and its beds have yielded an exceptionally detailed record of a warm climate biota for the Eocene. Its fossils include many plants, fish, insects, reptiles, amphibians and birds. The most significant are from a number of different mammal groups – bats, small horses, rodents and primates – and have exceptional preservation of skin, fur, muscles and stomach contents.

Not far away from Messel Pit is the Eocene-age Eckfeld Maar, where the fossils of small horses include a pregnant mare with foetus and placenta and have the remains of their last meal preserved in their stomachs.[36] Baruth Maar, also in Germany, is of similar size and age to Foulden, with laminated diatomite sediment, and is mainly known from cores. This locality was recognised initially from a geophysical anomaly: there is no surface expression of the maar, since it is completely covered by younger sediments.[37]

In Tanzania, Lake Mahenge, a small maar of Eocene age, has dolomitic sediment preserving plants, beetles, amphibians, bats and fish fossils. As yet only limited exposures have been studied, but it does represent one of very few pre-Quaternary Southern Hemisphere *Konservat-Lagerstätten*.[38]

Another important and highly diverse *Konservat-Lagerstätten*, closer in age to Foulden Maar, is the early to mid-Miocene Shanwang fossil biota in China. Fossil plants from Shanwang indicate that the vegetation was mixed deciduous and evergreen broad-leaved forest and the climate was humid warm-temperate to subtropical. Although average annual temperatures at Shanwang were similar to the present, the winters were much warmer and rainfall was less seasonal, allowing forest elements that preferred a warmer climate to grow there. As with Foulden, the macrofossil record seems to reflect plants growing near the maar and includes insect-pollinated taxa, while wind-dispersed pollen gives a more regional signal.[39]

These and various other maar *Lagerstätten* deposits occupy just a few square kilometres in total, but they contain an immense amount of information that is transforming our understanding of the history of life on Earth.

LEFT A few thousand years after its formation, Foulden Maar probably looked similar to this steep-sided maar crater at Mt Gambier in South Australia. Blue Lake is slightly smaller in diameter, with drier, more open forest vegetation covering the area once devastated by the volcanic eruption, rather than the subtropical rainforest that surrounded Foulden Maar. JGC

AN OUTSTANDING NATURAL FEATURE

The work at Foulden Maar has barely begun, and most collecting trips to the maar reveal new, often beautifully preserved fossils with important biogeographic and taxonomic implications. Much treasure still lies buried. At the time of writing, however, Foulden Maar still has no formal legal recognition as a site of international scientific interest, or any formal protection from future mining or other activities that would destroy it or prevent scientific research and community and educational access. No clear mechanism to remove the mining rights exists at present. Could it be transformed, like Messel Pit, from a mine site into a UNESCO World Heritage Site?

We suggest Foulden Maar should be given formal recognition and protection in district and regional plans and designated an Outstanding Natural Feature of International Significance. It is already listed as such in the Geoscience Society of New Zealand Geopreservation Inventory. The recommended definition of an Outstanding Natural Feature is 'a natural landform, physical system, or exposure of geological material that has outstanding geoscience, scenic/aesthetic, tourism, recreational, community and/or educational values or rarity'.[40]

Foulden Maar fits all these criteria: it is a natural landform; its geoscience significance is international; it is globally unique; it has high scenic and aesthetic values; its community and educational values are notable; its research potential is international in the fields of paleontology, paleoecology, climate research and geoethics; and its Indigenous cultural values as a taonga are important.

A VISION FOR FOULDEN MAAR

We consider Foulden Maar should be a publicly owned asset in the guardianship of Dunedin City Council and the Department of Conservation, and a model of sustainability and environmentally conscious stewardship. In an era of rapidly escalating climate change it is essential that Foulden Maar be maintained as a workable scientific site for continued climate and geological research, where students are welcome, where the public can access fossils as part of educational tours, and where issues such as geoethics are considered.

Foulden Maar has become a central research site for the current authors and for other New Zealand and international scientists who have come to appreciate the extraordinary treasure of fossil plants and animals present and the detailed environmental record contained in this ancient crater lake. The discoveries of Foulden Maar are a delight and a privilege to investigate, and give us new insights into life in Zealandia at the beginning of the Miocene. The quality of information available at the site has fostered the collaboration of systematic specialists, geochemists, geophysicists, paleoclimatologists and neoecologists.

Foulden Maar is an important educational asset. It provides a site for students to learn about the careful excavation and preparation of plant and animal fossils, the accurate use of stratigraphic techniques, the application of appropriate health and safety protocols, and the importance of working amicably with landowners and companies for the benefit of all.

Foulden Maar has taught us not to take for granted the preservation of key fossil sites in New Zealand. Most of these sites have no formal protection. Today, we rightly value modern threatened species and ecosystems but have few mechanisms for conserving sites occupied by our ancient biota. These represent the ancestors of our present plants and animals and other fossil organisms that lived in Zealandia for millions of years but could not persist when climates cooled. They are our biological heritage, and merit the same attention and respect as other aspects of our national heritage.

The rich and extraordinarily well-preserved fossil leaves, flowers, insects, spiders and freshwater fish of Foulden Maar have greatly advanced our knowledge of the history of New Zealand's unique terrestrial biota. The globally significant site contains a detailed

The main Foulden fossil-collecting team, 2017. Left to right: John Conran, Daphne Lee, Jennifer Bannister, Jon Lindqvist, Uwe Kaulfuss.

climate record and fossils that provide evidence for the role of the Southern Hemisphere in the biogeography and evolution of Earth's biota. Arguably, Foulden Maar is as important to our understanding of the origins of New Zealand's biota as the Messel Pit UNESCO World Heritage Site in Germany is for understanding the history of the biota of Europe. To lose such an important fossil site would be a tragedy both for New Zealand and for the wider paleontological community. It must be preserved.

APPENDIX

TABLE 1
PLANT MACROFOSSILS FROM FOULDEN MAAR AND THEIR APPARENT AFFINITIES

Updated from Lee et al., 'Biodiversity and palaeoecology of Foulden Maar: An early Miocene *Konservat-Lagerstätte* deposit in southern New Zealand', *Alcheringa* 40, 2016, pp. 525–41

TAXON	LEAVES/ STEMS	FLORAL PARTS	FRUITS/ SEEDS	ESTIMATED LIFE FORM	REFERENCE
FERNS					
Davalliaceae					
Davallia	frond			epiphyte	Conran, Kaulfuss, Bannister et al., 2010
Polypodiaceae					
Palaeosorum	frond			epiphyte	Kaulfuss, Conran, Bannister et al., 2019
GYMNOSPERMS					
Podocarpaceae					
Podocarpus	leaves		cone	tree	Pole, 1993d; Conran, Lee, Lee et al., 2014
Prumnopitys	leaves			tree	Lee, Conran, Lindqvist et al., 2012
BASAL ANGIOSPERMS					
Atherospermataceae					
Laurelia	leaves		fruits	tree	Conran, Bannister & Lee, 2013a
Lauraceae					
Beilschmiedia	leaves		fruits	tree	Bannister, Lee & Conran, 2012
Cryptocarya	leaves		fruits	tree	Bannister, Lee & Conran, 2012
Litsea	leaves	flowers	fruits	tree	Bannister, Lee & Conran, 2012
Monimiaceae					
Hedycarya	leaves	flowers	fruits	small tree	Conran, Bannister, Mildenhall et al., 2016
MONOCOTS					
Alstroemeriaceae					
Luzuriaga	leaves	flower		epiphyte, herb	Conran, Bannister, Mildenhall et al., 2014
Asteliaceae					
Astelia	leaves			epiphyte, herb	Maciunas, Conran, Bannister et al., 2011
Asparagaceae: subfam. Lomandroideae					
Cordyline	leaves			small tree	Conran, Bannister, Lee et al., 2015
Orchidaceae					
Dendrobium	leaves			epiphyte	Conran, Bannister & Lee, 2009
Earina	leaves			epiphyte	Conran, Bannister & Lee, 2009
Ripogonaceae					
Ripogonum	leaves, stem?			liane	Conran, Bannister, Lee et al., 2015
Typhaceae					
Typha	leaves		seeds	emergent macrophyte	Conran, Bannister, Lee et al., 2015

cont'd

EUDICOTS

TAXON	LEAVES/ STEMS	FLORAL PARTS	FRUITS/ SEEDS	ESTIMATED LIFE FORM	REFERENCE
Akaniaceae					
Akania		flowers		small tree	Conran, Kaulfuss, Bannister et al., 2019
Alseuosmiaceae					
Alseuosmia		flowers		shrub	Conran, Lee, Lee et al., 2014
Araliaceae					
Meryta	leaves			small tree	Lee, Kaulfuss, Conran et al., 2016
Pseudopanax	leaves		fruits	small tree	Conran, Lee, Lee et al., 2014
Schefflera	leaves*				
Bignoniaceae			seeds	tree or liane	Conran, Lee, Lee et al., 2014
Casuarinaceae					
Gymnostoma			fruit	small tree	Conran, Lee, Lee et al., 2014
cf. Celastraceae	leaves			tree or liane	Lee, Kaulfuss, Conran et al., 2016
Cunoniaceae					
cf. *Ackama* sp.	leaves	flowers		tree	Conran, Lee, Lee et al., 2014
Weinmannia	leaves		fruit	tree	Conran, Lee, Lee et al., 2014
Elaeocarpaceae					
cf. *Dubouzetia* sp.		flower*		shrub	
cf. *Elaeocarpus* sp.	leaves	flowers		tree	Conran, Lee, Lee et al., 2014
cf. *Sloanea* sp.	leaves	flowers	fruits	tree	Conran, Lee, Lee et al., 2014
Euphorbiaceae sens. lat.					
Malloranga	leaves	flowers		tree	Lee, Bannister, Raine et al., 2010
Euphorbiatheca			fruits, seeds	tree	Lee, Bannister, Raine et al., 2010
'Euphorb cf. *Baloghia*'		flowers	fruit	small tree	Conran, Lee, Lee et al., 2014
'Euphorb cf. *Scagaea*'		flower		small tree	Conran, Lee, Lee et al., 2014
Fabaceae (Leguminosae)	leaves?			shrub, small tree	Pole, 1996
subfam. Caesalpinioideae			fruit	small tree	Conran, Lee, Lee et al., 2014
Loranthaceae	leaves?	flower		hemiparasite	Conran, Lee, Lee et al., 2014
Malvaceae					
cf. subfam. Grewioideae	leaves			tree	Lee, Conran, Lindqvist et al., 2012
subfam. Sterculioideae	leaves*			tree	
Meliaceae					
'*Dysoxylum*'	leaves	flower	fruits	tree	Conran, Lee, Lee et al., 2014
Menispermaceae					
Hypserpa	leaves?		fruits	liane	Conran, Lee, Lee et al., 2014
Moraceae					
cf. *Streblus* sp.			fruits	small tree	Conran, Lee, Lee et al., 2014
Myrtaceae					
Metrosideros	leaves	flower*	fruits	tree or liane	Conran, Lee, Lee et al., 2014
Syzygium	leaves		seeds	tree	Conran, Lee, Lee et al., 2014

TAXON	LEAVES/ STEMS	FLORAL PARTS	FRUITS/ SEEDS	ESTIMATED LIFE FORM	REFERENCE
Onagraceae					
Fuchsia		flower, anthers		tree	Lee, Bannister, Kaulfuss et al., 2013
Pittosporaceae					
cf. *Pittosporum* sp.			fruit?	shrub, small tree	Conran, Lee, Lee et al., 2014
Polygalaceae					
cf. *Muehlenbeckia* sp.		flower*		liane	
Primulaceae (including Myrsinaceae)					
Myrsine	leaves, stem			tree	Lee, Conran, Lindqvist et al., 2012
Proteaceae					
cf. *Gevuina* sp.	leaves			tree	Carpenter, Bannister, Lee et al., 2012
cf. *Alloxylon* sp.	leaves			tree	Carpenter, Bannister, Lee et al., 2012
Rubiaceae					
cf. *Coprosma* sp.			fruits?	shrub, small tree	Conran, Lee, Lee et al., 2014
Rutaceae			seeds?		Conran, Lee, Lee et al., 2014
Fouldenia		flower		tree?	Bannister, Lee & Raine, 2005
cf. *Neoschmidia* sp.		flower		tree	Conran, Lee, Lee et al., 2014
Melicope	leaves	flower?		tree	Conran, Lee, Lee et al., 2014
Sapindaceae					
Alectryon	leaves			tree	Lee, Conran, Lindqvist et al., 2012
Cupaniopsis	leaves			tree	Lee, Conran, Lindqvist et al., 2012
Sapotaceae					
cf. *Planchonella/Pouteria*		flower bud		tree	Conran, Lee, Lee et al., 2014
Winteraceae	leaves			shrub, small tree	Pole, 1996

* new/additional unpublished data presented here

TABLE 2
ANIMAL MACROFOSSILS FROM FOULDEN MAAR

Updated from Lee et al., 'Biodiversity and palaeoecology of Foulden Maar: An early Miocene *Konservat-Lagerstätte* deposit in southern New Zealand', *Alcheringa 40*, 2016, pp. 525–41

TAXON	PART PRESERVED	REFERENCE
FISH		
Galaxiidae *Galaxias effusus*	articulated body fossils	Lee, McDowall & Lindqvist, 2007; Kaulfuss, Lee, Robinson et al., 2020
Anguillidae cf. *Anguilla* sp.	articulated body fossil	Kaulfuss, 2013
BIRDS		
Various waterbirds	coprolites	Lindqvist & Lee, 2009
SPIDERS		
Arachnida		
Incertae sedis 1	articulated (pyritised)	Selden & Kaulfuss, 2019
Incertae sedis 2	articulated (pyritised)	Selden & Kaulfuss, 2019
Araneomorphae		
Incertae sedis	articulated	Selden & Kaulfuss, 2019
Mygalomorphae		
Idiopidae	articulated	Selden & Kaulfuss, 2019
INSECTS		
Blattaria: Isoptera Stolotermitidae *Stolotermes kupe*	isolated wing	Kaulfuss, Harris & Lee, 2010
Kalotermitidae		
Waipiatatermes matatoka	isolated wing	Engel & Kaulfuss, 2016
Taieritermes krishnai	isolated wing	Engel & Kaulfuss, 2016
Otagotermes novazealandicus	isolated wing	Engel & Kaulfuss, 2016
Pterotermopsis fouldenica	isolated wing	Engel & Kaulfuss, 2016
Hemiptera Aradidae: *Aneurus* sp.	articulated	Kaulfuss, Wappler, Heiss et al., 2011
Cixiidae	articulated; isolated wing	Kaulfuss, Lee, Barratt et al., 2014
Coccidae	scale attached in life position	Kaulfuss, Lee, Barratt et al., 2014
Diaspididae	scales attached in life position	Harris, Bannister & Lee, 2007

TAXON	PART PRESERVED	REFERENCE
Cicadellidae	isolated hindwing	Herein
Tingidae	articulated	Herein
Coleoptera Cerambycidae	articulated	Kaulfuss, Lee, Barratt et al., 2014
Chrysomelidae	articulated	Kaulfuss, Lee, Barratt et al., 2014
Curculionidae	articulated	Kaulfuss, Lee, Barratt et al., 2014
Elateridae	articulated	Herein
Hydrophilidae	articulated	Kaulfuss, Lee, Barratt et al., 2014
Nitidulidae	articulated	Herein
Staphylinidae *Sphingoquedius meto*	articulated	Jenkins Shaw, Solodovnikov, Ming Bai & Kaulfuss, 2020
Pselaphinae *Sagola* sp.	articulated	Kaulfuss, Lee, Barratt et al., 2014
Osoriinae *Nototorchus* sp.	articulated	Kaulfuss, Lee, Barratt et al., 2014
Hymenoptera Formicidae		
Ambyloponinae	articulated	Kaulfuss, Harris, Conran et al., 2014
Ponerinae *Austroponera schneideri*	articulated	Kaulfuss & Dlussky, 2015
Formicinae *Myrmecorhynchus novaeseelandiae*	articulated	Kaulfuss & Dlussky, 2015
Ectatomminae *Rhytidoponera gibsoni* *Rhytidoponera waipiata*	articulated articulated	Kaulfuss & Dlussky, 2015 Kaulfuss & Dlussky, 2015
Ichneumoidea	articulated	Kaulfuss, Lee, Barratt et al., 2014
Chalcidoidea	articulated	Kaulfuss, Lee, Barratt et al., 2014
Trichoptera Leptoceridae	larvae and larval cases	Herein
Oeconesidae	larvae and larval cases	Herein
Diptera Chironomidae	pupae	Herein
Chaoboridae	pupae	Herein
Tipuloidae	articulated	Kaulfuss, Lee, Barratt et al., 2014
Empididae	articulated	Herein

TABLE 3
COMPARISON BETWEEN FOULDEN MAAR AND SOME OTHER CENOZOIC MAARS

MAAR	LOCATION	AGE	DIAMETER (M)	SEDIMENT	SURFACE/ CORED
Foulden	New Zealand	early Miocene	1000	diatomite	surface exposures; 183m core
Hindon	New Zealand	middle Miocene	c. 500×800	diatomite; carbonaceous mudstone	surface exposures; 40m core
Messel	Germany	middle Eocene	>2000	oil shale	surface exposures; cores
Eckfeld	Germany	middle Eocene	1300	oil shale	surface exposures; cores
Mahenge	Tanzania	Eocene	400	dolomitised mudstone and siltstone	surface exposures to 3.8m
Baruth	Germany	Oligocene	c. 1000	diatomite	cores
Enspel	Germany	Oligocene	1300×1700	oil shale	surface exposures; cores to 250m +
Randeck	Germany	middle Miocene	c. 1200	limestone; bituminous laminites	surface exposures; cores
Shanwang	China	middle Miocene	c. 100	diatomaceous shale; mudstone	surface exposures; cores
Camp dels Ninots	Spain	Pliocene	650×400	laminated clays; sandstone; diatomite	surface exposures; cores

PLANTS	ARTHROPODS	VERTEBRATES	REFERENCE
Lauraceae-dominant mixed mesothermal rainforest	10+ orders; 21+ families; 4 spiders	fish	Lindqvist & Lee (2009); Lee et al. (2012); Kaulfuss et al. (2014b)
Nothofagaceae-dominant mixed mesothermal forest	5+ orders	fish	Kaulfuss & Moulds (2015) Kaulfuss et al. (2018)
mixed angiosperm and conifer mesophytic rainforest	12+ orders: diverse families; beetle-dominated	fish, amphibians, reptiles, birds, mammals	Schaal & Ziegler (1992); Selden & Nudds (2012); Büchel & Wuttke (2014)
mixed angiosperm and conifer mesophytic rainforest	11 orders, >4600 specimens; diverse families; beetle-dominated	fish, amphibians, reptiles, birds, mammals	Wilde & Frankenhäuser (1998); Lutz & Kaulfuss (2006); Lutz et al. (2010)
Leguminosae-dominant subtropical woodland	beetles	amphibians, fish, 1 microbat	Kaiser et al. (2006)
mixed angiosperm and conifer mesophytic forest	beetles		Goth & Suhr (2007)
warm-temperate, mixed mesophytic forest	14+ orders, >5000 specimens	fish, amphibians, reptiles, birds, mammals	Poschmann et al. (2010); Schindler & Wuttke (2010); Wuttke et al. (2010; 2015)
168 taxa; subhumid sclerophyllous and mixed mesophytic forests	15 orders; 79 taxa	fish, amphibians, reptiles, birds, mammals	Rasser et al. (2013)
127 taxa; mixed mesophytic forests	12 orders; 80 families	fish, amphibians, reptiles, birds, mammals	Liu & Leopold (1992); Yang & Yang (1994); Sun et al. (2002)
mixed angiosperm and conifer mesophytic forests		fish, amphibians, reptiles, birds, mammals	Gómez de Soler et al. (2012) Bolós et al. (2021)

TABLE 4
LIST OF TAXA FOR WHICH FOULDEN MAAR IS THE TYPE LOCALITY

Foulden Maar is currently the type locality for 27 species of plants and pollen, 10 species of insects and one species of fish. These taxa are all endemic to New Zealand.

PLANTS (INCLUDING DIATOMS AND NOVEL POLLENS)

Encyonema jordaniforme Krammer 1997

Fouldenia staminosa Bannister, D.E. Lee & Raine 2005

Palaeosorum waipiata U. Kaulfuss, Conran, Bannister, Mildenhall & D.E. Lee 2019

Davallia walkeri Conran, U. Kaulfuss, Bannister, Mildenhall & D.E. Lee 2010

Podocarpus travisiae Pole 1993

Dendrobium winikaphyllum Conran, Bannister & D.E. Lee 2009

Earina fouldenensis Conran, Bannister & D.E. Lee 2009

Laurelia otagoensis Conran, Bannister & D.E. Lee 2013

Laurophyllum maarensis Bannister, Conran & D.E. Lee 2012

Laurophyllum microphyllum Bannister, Conran & D.E. Lee 2012

Laurophyllum waipiata Bannister, Conran & D.E. Lee 2012

Laurophyllum taieriensis Bannister, Conran & D.E. Lee 2012

Laurophyllum calicarioides Bannister, Conran & D.E. Lee 2012

Laurophyllum sylvestris Bannister, Conran & D.E. Lee 2012

Laurophyllum lacustris Bannister, Conran & D.E. Lee 2012

Laurophyllum vulcanicola Bannister, Conran & D.E. Lee 2012

Laurophyllum otagoensis Bannister, Conran & D.E. Lee 2012

Litseopsis nova-zelandiae Bannister, Conran & D.E. Lee 2012

Malloranga fouldenensis D.E. Lee, Bannister, J.I. Raine & Conran 2010

Euphorbiotheca mallotoides D.E. Lee, Bannister, J.I. Raine & Conran 2010

Astelia antiquua Maciunas, Conran, Bannister, R. Paull & D.E. Lee 2011

Euproteaciphyllum alloxylonoides R.J. Carpenter, Bannister, D.E. Lee & G.J. Jordan 2012

Euproteaciphyllum pacificum R.J. Carpenter, Bannister, D.E. Lee & G.J. Jordan 2012

Fuchsia antiqua D.E. Lee, Conran, Bannister, U. Kaulfuss & Mildenhall 2013

Luzuriaga peterbannisteri Conran, Bannister, Mildenhall & D.E. Lee 2014

Liliacidites contortus Mildenhall & Bannister 2014

Dicyclopsodites leei Mildenhall, E.M. Kennedy & J.G. Prebble 2014

Akania gibsonorum Conran, U. Kaulfuss, Bannister, Mildenhall, & D.E. Lee 2019

INSECTS

Stolotermes kupe Kaulfuss, Harris & Lee 2010

Waipiatatermes matatoka Engel & Kaulfuss 2016

Taieritermes krishnai Engel & Kaulfuss 2016

Otagotermes novazealandicus Engel & Kaulfuss 2016

Pterotermopsis fouldenica Engel & Kaulfuss 2016

Rhytidoponera waipiata Kaulfuss & Dlussky 2015

Rhytidoponera gibsoni Kaulfuss & Dlussky 2015

Myrmecorhynchus novaeseelandiae Kaulfuss & Dlussky 2015

Austroponera schneideri Kaulfuss & Dlussky 2015

Sphingoquedius meto Jenkins Shaw, Solodovnikov, Ming Bai & Kaulfuss 2020

FISH

Galaxias effusus Lee, McDowall & Lindqvist 2007

TABLE 5
FOULDEN MAAR TIMELINE 1875–2021

DATE	EVENT	AUTHOR	COMMENTS
1875	First mention in print of diatomaceous earth deposit at Strath Taieri	Hutton & Ulrich: *Report on the Geology and Gold Fields of Otago*	
1884	Foulden Hill subdivided off Cottesbrook Run by Thompson family		
1910 4 Jul	Diatomaceous earth deposit estimated to cover 120 acres (48ha); 385 bags sent to England	Report in *ODT*	
1941	79 tons diatomite sold		
1942	74 tons diatomite sold		
1954	Unpublished mining report prepared on 'The Middlemarch diatomite deposit. The completion of the exploratory programme and development'.	Gordon, 1954	
1959 1–3 Sep	Papers presented at Fourth Triennial Mineral Conference	Gordon, 1959	School of Mines and Metallurgy, Dunedin
1960	Plant microfossils mentioned; thought to be Pliocene	Couper, 1960	
1962–65	MSc thesis included Foulden Hill; fish and leaf fossils reported	Travis, 1965	University of Otago, unpublished thesis
1966	1:250,000 geological map published	McKellar, 1966	
1970–2019	Staff and students from Geology Department visited on field trips		
1993	First leaf fossil described	Pole, 1993	
1996	Plant macrofossils described	Pole, 1996	
2000 4 Jul	'Consent granted for Featherston Resources diatomite mine which will employ up to 30 people in 5 years'	Report in *ODT*	
2003 Nov	Field trip guide for Geoscience Society of New Zealand (GSNZ) Dunedin conference	Lee, Lindqvist, Douglas, Bannister, Cieraad	GSNZ miscellaneous publication
2005	First fossil flower described	Bannister, Lee, Raine	*NZ Journal of Botany*
2006 9–13 Jul	Geophysics Field School		University of Otago
2007	Fossil *Galaxias* described	Lee, McDowall, Lindqvist	
2007	Fossil scale insects described	Harris, Bannister, Lee	

cont'd

DATE	EVENT	AUTHOR	COMMENTS
2008–11	Marsden grant awarded	Lee, Wilson, Gorman et al.	
2009	First paleoclimate paper	Lindqvist & Lee	
2009	Hons thesis on fossil Asteliaceae	Macunias, 2009	University of Adelaide, unpublished thesis
2009 10 Jun–3 Jul	Marsden drilling programme, 184m core retrieved		
2009 Nov	Field trip guide for GSNZ Ōamaru conference	Lee, Lindqvist, Mildenhall, Bannister, Kaulfuss	GSNZ miscellaneous publication
2009–	Publications on plant and insect fossils	Many authors, ongoing	
2011–14	Marsden grant awarded	Lee, Mildenhall et al.	
2012	MSc thesis on geophysical characterisation of maar	Jones, 2012	University of Otago, unpublished thesis
2013 Jan	Field trip to Foulden Maar for Southern Connection international conference, Dunedin		Lee, Conran, Kaulfuss field trip leaders
2013	PhD thesis on geology and paleontology of Foulden Maar	Kaulfuss, 2013	University of Otago, unpublished thesis
2014	PhD thesis on climate change at the Oligocene–Miocene boundary	Fox, 2014	University of Otago, unpublished thesis
2014–18	Marsden grant awarded	Lee, Mildenhall et al.	
2014 1 Nov	Overseas mining company purchases 42ha with Overseas Investment Office (OIO) approval	'Diatomite mine sold for $650k': article by reporter Simon Hartley in *ODT*	
2015	PhD thesis on Miocene terrestrial climate from plant fossils	Reichgelt, 2014	University of Otago, unpublished thesis
2016	Hons thesis on fossil Ripogonaceae	Kerr, 2016	University of Adelaide, unpublished thesis
2017 4 Feb	Foulden Hill property advertised for sale		
2018 5 Jul	Presentation to Strath Taieri Community Board by new mining company	Reported in *ODT*	
2018 2 Jun	'Miners, scientists unlikely allies'	Article by reporter Simon Hartley in *ODT*	
2018 28 May	'Potential to create 100 jobs'	Article by reporter Simon Hartley in *ODT*	
2018 18 Jun	'Funding secured by diatomite mine firm'	Article by reporter Simon Hartley in *ODT*	Seed funding of $US20m secured from Goldman Sachs
2018 18 Jun	'Company to begin meetings on quarry'	Article by reporter Simon Hartley in *ODT*	

DATE	EVENT	AUTHOR	COMMENTS
2019 20 Apr	'Leaked report sheds light on mine project'	Article by reporter Simon Hartley in *ODT*	Confidential report from Goldman Sachs
2019 9 May	Community meeting in Middlemarch		
2019 18 May	Cr Aaron Hawkins, Dunedin City Council (DCC), supports Foulden Maar preservation		
2019 19 May	'Turning our Taonga into pet food'	Article by Paul Little in *NZ Herald*	
2019 20 May	'Digging for the truth on fossils, profit, and the Foulden Maar mine'	Article by Juliet Gerrard republished in *The Spinoff*	
2019 24 May	'Foulden Maar: Unjustifiable vandalism and grand promises'	Article by Rod Oram on *Newsroom*	
2019 25 May	'Sifting through layers of history'	Article by reporter Elena McPhee in *ODT*	
2019 25 May	Nic Rawlence presents talk for Daphne Lee at Wise Response forum, Otago University		
2019 28 May	Daphne Lee gives invited talk to DCC on scientific importance of Foulden Maar		
2019 8 Jun	'Mining of Foulden Maar described as "obscene"'	Article by reporter Elena McPhee in *ODT*	
2019 11 Jun	'Plaman seeks to be party to Environment Court appeals'	Article by reporter Simon Hartley in *ODT*	
2019 12 Jun	GSNZ revises OIO submission: 'Geoscience Society fears loss of fossils'	Article by reporter Tim Miller in *ODT*	
2019 13 Jun	Plaman Resources placed in liquidation		
2019 14 Jun	'DCC formally opposed to expansion of mining site'	Article by reporter Tim Miller in *ODT*	
2019 17 Jun	Call for fossils to be saved	Article by Anthony Harris in *ODT*	
2019 22 Jun	'Spotlight on Plaman Resources'	Article by reporter Simon Hartley in *ODT*	
2019 25 Jun	Daphne Lee gives invited talk to Middlemarch community		
2019 25 Jun	Daphne Lee gives invited talk at Otago Museum		
2019 26 Jun	'Further bad news for troubled Plaman'	Article by reporter Simon Hartley in ODT	

cont'd

DATE	EVENT	AUTHOR	COMMENTS
2019 26 Jun	Public meeting in Dunedin indicates support for maar protection	Article by reporter Chris Morris in *ODT*	
2019 31 Jul	Plaman withdraws application to OIO to buy 42ha of land near Middlemarch	Report by Land Information New Zealand	'The OIO received 42 submissions … about the application, including information about the significant fossil site on the land.'
2019 1 Aug	'Plaman withdraws; maar still vulnerable'	Article by reporter Elena McPhee in *ODT*	
2019 3 Aug	'Foulden Maar battle has been won; war not over'	Letter to the Editor, Shane Loader, *ODT*	
2019 9 Nov	'Boost for Foulden Maar fans'	Editorial in *ODT*	
2020 2 Feb	DCC poised to buy fossil deposit land	DCC Chief Executive Sue Bidrose, *ODT*	
2020 23 Jul	'DCC working with receivers, no comment on Foulden Maar'	Article by Hamish Maclean in *ODT*	
2020 21 Aug	New paper on Foulden Maar in Climate of the Past results in global interest	Tammo Reichgelt et al., 2020	Media coverage by BBC, *Stuff*, *ODT* etc

GLOSSARY

achene: single-seeded, dry, non-splitting fruit

alate: winged female or male insect that pairs up during nuptial flight and sheds its wings shortly afterwards

anther: part of a stamen that contains the pollen

anoxic: depleted or lacking in dissolved oxygen

arillate: a seed with a fleshy appendage

arthropods: largest phylum in the animal kingdom that includes insects and spiders: all have a hard jointed exoskeleton and paired jointed legs

basal angiosperm: flowering plant that separated early on from the lineage leading to most modern flowering plants and pollen with a single aperture

basalt: fine-grained, dark-coloured volcanic rock made up mostly of feldspar and pyroxene; the most common type of volcanic rock

basement rock: the oldest rock unit in a region; generally igneous or metamorphic rock beneath layers of younger sediment

bedding: layers of sedimentary rock of varying thickness and character

bedding plane: surface in a sedimentary rock parallel to the original surface of deposition

benthic: relating to or occurring at the bottom of a body of water

biogenic: produced by living organisms

biogenic silica: amorphous (non-crystalline) silica in sediments derived from biological sources such as diatoms and sponges

biogeography: the study of the distribution of species and ecosystems in geographic space and through geologic time

biostratigraphy: the study of the relative age of sedimentary layers using fossils

biota: plant and animal life of a region

breccia: sedimentary rock made up of angular coarse fragments

Cenozoic: period of Earth's history from 66 Ma to the present: includes the Paleogene, Neogene and Quaternary

clade: a group of biological taxa (such as species) that includes all descendants of one common ancestor

clast: fragment of rock broken off from a larger rock

core: sample obtained by drilling into sediment or rock

corolla: the collective term for the petals – the inner ring of modified leaf-like, usually coloured structures in a flower designed to advertise to pollinators

cotyledon: the first leaf-like structures to emerge from a seed at germination

cuticle: waxy layer that covers the leaves of all plant species

diatom: microscopic single-celled alga with a siliceous shell or frustule

diatomite: silica-rich sedimentary rock made up mostly of diatoms

diatreme: steep-sided, typically cone-shaped structure enclosed by country rock and with a volcaniclastic infill

dicot: any flowering plant that has two cotyledons in its seed (cf. monocot)

dike: tabular body, usually of igneous rock, which cuts across the structure of adjacent rocks

dip: angle between a horizontal and an inclined plane

domatia: tiny pockets or hair-tufts in the main veins of a leaf produced by the plant to provide a home for arthropods

drupe, drupelet: single-seeded fleshy fruit with a stone-like inner wall around the seed

ecology: the study of the relationships of living things to each other and their environment

ecosystem: community of interacting organisms and their environments

elytra: hardened forewings of a beetle

endemic: native and restricted to a certain place

endocarp: hard seed-like fruit

ENSO cycle: El Niño–Southern Oscillation periodic cycle affecting tropical and subtropical climates: sea surface temperatures warm during El Niño and cool during La Niña phases

epiphyllous: growing on leaves

epiphyte: organism that uses a plant as a substrate on which to live but which is not parasitic on the plant

eudicot: a lineage of flowering plants that has two seed leaves after seed germination and pollen with three or more apertures

eusocial: relating to insects such as ants or termites that live in cooperative groups

extinction: disappearance of an entire class, family, genus or species of plant or animal

fault: major fracture in rocks along which movement has taken place

fauna: the animal life in a particular region or geological time interval

feldspar: a group of light-coloured potassium, sodium, calcium and aluminium silicate minerals – the most common type of minerals in the Earth's crust

flora: plant life of a particular region or geological time interval

fluvial: relating to rivers and streams

foliation: tendency of a metamorphic rocks to split along flat planes

food web: complex set of feeding interactions and energy flows between species in an ecosystem

foraminifera: single-celled protists found in all marine environments and in marine sediments

formation: body of rock unified by origin, age or composition

fossil: remains, trace or imprint of a plant or animal preserved in rock

Fossil-Lagerstätten: exceptionally well-preserved fossil biotas

frustule: tiny, hard, box-like cell wall of a diatom made of opaline silica

genus (pl. genera): in Linnaean classification, the second-lowest level used to classify organisms, and consisting of one or more related species – for example, *Galaxias*, the genus of fish to which whitebait belong

geomagnetic polarity: configuration of the north and south magnetic poles of the Earth's geomagnetic field

geophysics: the study of the physical properties of the Earth

gravity measurement: geophysical method based on measurement of the attractive force of the Earth that varies with altitude, latitude and the density of local rock masses

groundwater: subsurface water contained in pores and fissures in rock beneath the soil, most of which is below the water table

habitat: natural home of an organism

Hindon Maar Complex: four small maar craters in farmland near Hindon

holotype: type specimen of a species

igneous: describing rock that is formed by the cooling and consolidation of molten magma

inflorescence: connected group of flowers

invertebrate: animal that has no backbone (e.g. insects, molluscs)

Konservat-Lagerstätten: exceptional fossil deposits where peculiar preservational conditions have allowed even the soft tissue of animals and plants to be preserved

lacustrine: relating to lakes

lithophyte: plant that grows on rocks

littoral: shallow, near-shore part of a lake or the sea

Ma: abbreviation for millions of years ago

maar: type of volcano with a crater cut into the pre-eruptive landscape, surrounded by a tephra ring, and underlain by a diatreme

macrofossil: fossil, typically more than 1mm across, that is visible to the naked eye

macrophyte: large, multicellular plant or alga growing in a lake, either floating or attached to an object

magma: molten rock generated within the Earth

magnetism: magnetic property of rocks or minerals

marine transgression: inundation of the land by the sea, caused by rising sea level or subsiding land

megascleres, **microscleres** and **gemmuloscleres**: types of spicules, the structural elements found in sponges

meromictic: having a well-mixed oxygenated upper water layer and a lower, unmixed water column that is depleted of oxygen

metamorphic rock: changed by heat and pressure, often at depth within the Earth's crust

microfossil: fossil, typically less than 1mm across, that can only be studied under a microscope

micron: 1 micron, written 1μm, is one-thousandth of a millimetre

Miocene: period of Earth's history from 23.03 to 5.3 Ma

molecular phylogeny: the study of the evolutionary relationships of an organism

monocot: flowering plant with a single embryonic leaf

morphology: form of a plant or animal

mucilaginous: thick and gluey/sticky

nacreous: lustrous or pearlescent

Neogene: period of Earth's history from 23.03 to 2.58 Ma; includes the Miocene and Pliocene

Nothofagus: scientific genus name for southern beech

oil shale: very fine-grained rock with high levels of hydrocarbons

opaline silica: non-crystalline biogenic silica

outcrop: exposure of rock at the surface

overburden: overlying rock or soil that is removed before open-cast mining

paleoclimate: ancient climate as determined from the geological record

Paleogene: period of Earth's history from 66 to 23.03 Ma; includes the Paleocene, Eocene and Oligocene

paleolake: fossil lake such as Lake Manuherikia

paleontology: the study of life in the geological past

palynology: the study of spores and pollen and other acid-resistant organic structures

paniculate: much-branched connected cluster of flowers

pennate: boat-shaped

perithecia: fungal fruiting bodies characterised by being rounded or flask-like and releasing their spores through a pore-like hole

petiole: stalk holding a leaf base to its stem

photomicrograph: photograph taken through a microscope lens

phylogeny: sequence of events involved in the evolution of a species, genus, etc.

phytoplankton: microscopic, photosynthetic aquatic algae, generally buoyant and found floating in the upper part of the water column in lakes and oceans

pinnate: with leaflets growing opposite each other in pairs

planktic: living on or near the water surface

pollen: microscopic male reproductive cells of plants, dispersed from the anthers

pollinium: pollen mass

pronotum: plate-like cover of the first segment of an insect's thorax

protist: a diverse group of microscopic, mostly unicellular organisms that possess a nucleus in their cells but are not classed as plants, animals or fungi

pyrite: FeS_2, a common iron sulfide with a metallic lustre and brassy yellow colour, also known as 'fool's gold'

pyroclastic: ejected and deposited during or after volcanic eruptions

Quaternary: period of Earth's history from 2.6 Ma to the present: includes the Pleistocene (the period of the Ice Ages) and Holocene

quartz: a commonly occurring mineral composed of silicon dioxide (SiO_2)

radiometric dating: method for determining the age of rocks using radioactive isotopes

schist: metamorphic rock that splits along layers of mica minerals

sclerophyll: vegetation with tough, often small leaves adapted to low rainfall or seasonally hot, dry environments

sedimentary: formed from sediments, whether laid down by rivers, lakes, seas, wind or glaciers

SEM: scanning electron microscope

siliceous: consisting of silica, silicon dioxide (SiO_2)

sorus (pl. sori): cluster of spore sacs on a fern frond

spicule: elongate structural elements that support the soft tissue of a sponge

sporangium (pl. sporangia): stalked, rounded organs containing spores

spore: asexual reproductive cell produced by bacteria, fungi, algae and plants that can grow directly into a new individual

stoma (pl. stomata): pore in a leaf through which gas exchange takes place

subrounded: not quite circular

taphonomy: the study of how plants or animals become fossilised and preserved in the fossil record

taxon (pl. taxa): general term for a taxonomic group such as family, genus or species

tephra: loose fragmented material from a volcanic eruption

tephra ring: low rim of volcanic debris surrounding a maar crater

terrestrial: land-based

thin section: very thin slice of rock that can be analysed under a petrological microscope

thorax: central part of an insect's body to which its legs and wings are attached

tribe: a taxonomic category that ranks above genus and below family or subfamily

tricolpate: with three apertures

tuff: lithified volcanic ash

turbidite: sediments deposited from a low-density mix of sediment and water

type locality: place where an original specimen was collected

type specimen: original specimen from which a description of a new species is made

varve: a single year's worth of sediment; thin bands of sediment deposited as couplets, each layer differing slightly in thickness and colour

volcanic field: area of localised volcanic activity, containing clusters of many volcanoes

vomer: part of the roof of the mouth of an eel that holds the teeth, just behind the upper jaw

NOTES

ABBREVIATIONS

AmJB American Journal of Botany
AuJB Australian Journal of Botany
AuSB Australian Systematic Botany
BJLS Botanical Journal of the Linnean Society
BR Botanical Review
GPC Global and Planetary Change
JRSNZ Journal of the Royal Society of New Zealand
NZJE New Zealand Journal of Ecology
NZJGG New Zealand Journal of Geology and Geophysics
PE Palaeontologia Electronica
PPEES Perspectives in Plant Ecology, Evolution and Systematics
PPP500 Palaeogeography, Palaeoclimatology, Palaeoecology
RPP Review of Palaeobotany and Palynology
SB Systematic Biology

INTRODUCTION

1. Time scale adapted by Jenny Stein and Stephen Read from information in J.I. Raine, A.G. Beu, A.F. Boyes, H.J. Campbell, R.A. Cooper, J.S. Crampton, M.P. Crundwell, C.J. Hollis, H.E.G. Morgans & N. Mortimer, 'New Zealand Geological Timescale NZGT 2015/1', *New Zealand Journal of Geology and Geophysics* 58, 2015, pp. 398–403.
2. P. Selden & J. Nudds, *Evolution of Fossil Ecosystems* (2nd edn) (London: Manson Publishing, 2012).
3. U. Kaulfuss, D.E. Lee, J.-A.Wartho et al., 'Geology and palaeontology of the Hindon Maar Complex: A Miocene terrestrial fossil *Lagerstätte* in southern New Zealand', *PPP 500*, 2018, pp. 52–68; D.E. Lee, U. Kaulfuss, J.G. Conran et al., 'Biodiversity and palaeoecology of Foulden Maar: An early Miocene *Konservat-Lagerstätte* deposit in southern New Zealand', *Alcheringa 40*, 2016, pp. 525–41.

CHAPTER 1

1. P.J. Forsyth (ed.), *Geology of the Waitaki Area 1:250 000 geological map 19* (Lower Hutt: Institute of Geological & Nuclear Sciences, 2001).

2. Ibid.
3. C. Travis, 'Geology of the Slip Hill area east of Middlemarch', MSc thesis, University of Otago, Dunedin, 1965.
4. D.S. Coombs, C.J. Adams, B.P. Roser & A. Reay, 'Geochronology and geochemistry of the Dunedin Volcanic Group, eastern Otago, New Zealand', *NZJGG 51*, 2008, pp. 195–218.
5. J.M. Scott, A. Pontesilli, M. Brenna et al., 'The Dunedin Volcanic Group and a revised model for Zealandia's alkaline intraplate volcanism', *NZJGG 63*, 2020, pp. 510–29.
6. Coombs, Adams, Roser & Reay, 'Geochronology and geochemistry of the Dunedin Volcanic Group'; Ibid.
7. U. Kaulfuss, 'Geology and paleontology of Foulden Maar, Otago, New Zealand', PhD thesis, University of Otago, Dunedin, 2013, Fig. 2.8; U. Kaulfuss, 'Crater stratigraphy and the post-eruptive evolution of Foulden Maar, southern New Zealand', *New Zealand Journal of Geology and Geophysics 60*, 2017, pp. 410–32.
8. B.W. Hayward, *Volcanoes of Auckland: A field guide* (Auckland: Auckland University Press, 2019b).
9. V. Lorenz, 'Maar-diatreme volcanoes, their formation, and their setting in hard-rock or soft-rock environments', *GeoLines 15*, 2003, pp. 72–83.
10. J.D.L. White & P.S. Ross, 'Maar-diatreme volcanoes: A review', *Journal of Volcanology and Geothermal Research 201*, 2011, pp. 1–29.
11. J.K. Lindqvist & D.E. Lee, 'High-frequency paleoclimate signals from Foulden Maar, Waipiata Volcanic Field, southern New Zealand: An early Miocene varved lacustrine diatomite deposit', *Sedimentary Geology 222*, 2009, pp. 98–110.
12. D.A. Jones, 'The geophysical characterisation of the Foulden Maar', MSc thesis, University of Otago, Dunedin, 2012: http://hdl.handle.net/10523/2334; D.A. Jones, G.S. Wilson, A.R. Gorman et al., 'A drill-hole calibrated geophysical characterisation of the 23 Ma Foulden Maar stratigraphic sequence, Otago, New Zealand', *NZJGG 60*, 2017, pp. 465–77.

13. U. Kaulfuss, 'Crater stratigraphy and the post-eruptive evolution of Foulden Maar, southern New Zealand', *NZJGG 60*, 2017, pp. 410–32.

14. B.R.S. Fox, G.S. Wilson & D.E. Lee, 'A unique annually laminated maar lake sediment record shows orbital control of Southern Hemisphere mid-latitude climate across the Oligocene–Miocene boundary', *GSAB 128*, 2016, pp. 609–26.

15. D.E. Lee, U. Kaulfuss, J.G. Conran et al., 'Biodiversity and palaeoecology of Foulden Maar: An early Miocene *Konservat-Lagerstätte* deposit in southern New Zealand', *Alcheringa 40*, 2016, pp. 525–41.

16. The cross-section of the maar is based on the schematic model for a maar diatreme proposed by Lorenz (2003) and White & Ross (2011). Modified from Jones et al., 2017, Fig. 9.

17. C. Timms, personal communication, 2007.

18. B.R.S. Fox, 'Climate change at the Oligocene/Miocene boundary', PhD thesis, University of Otago, Dunedin, 2014.

19. Kaulfuss, 'Crater stratigraphy and the post-eruptive evolution of Foulden Maar, southern New Zealand'.

CHAPTER 2

1. A.G. Hocken, 'The early life of James Hector, 1834 to 1865: The first Otago provincial geologist', PhD thesis, University of Otago, Dunedin, 2008.

2. F.W. Hutton & G.H.F. Ulrich, *Report on the Geology and Gold Fields of Otago* (Dunedin: Provincial Council Otago, 1875).

3. Ibid.

4. Graeme Thompson, personal communication, 29 April 2019.

5. *US Geological Survey, Mineral Commodity Summaries, January 2021* (US Geological Survey, Reston, VA, 2021): https://pubs.usgs.gov/periodicals/mcs2021/mcs2021-diatomite.pdf

6. P.G. Morgan, 'Minerals and mineral substances of New Zealand', *NZ Geological Survey Bulletin 3*, Wellington: Government Printer, 1927).

7. L.I. Grange, 'Diatomite: Principal New Zealand sources and uses', *New Zealand Journal of Science and Technology 12*, 1930, pp. 94–99.

8. F.R. Gordon, 'The Middlemarch Diatomite Deposit: The completion of the exploratory programme and development' (unpublished mining report) (School of Mines and Metallurgy, University of Otago, Dunedin, 1954); F.R. Gordon, 'Paper 136: The occurrence of diatomite near Middlemarch, Otago', in *Proceedings of the Fourth Triennial Mineral Conference, School of Mines and Metallurgy, University of Otago, Dunedin, New Zealand, 1–3 September 1959*, (University of Otago, Dunedin, 1959a), pp. 9; F.R. Gordon, 'Paper 137: The properties and uses of Middlemarch diatomite', in ibid., p. 13.

9. B. Thompson, R.L. Brathwaite & A.B. Christie, *Mineral Wealth of New Zealand*, IGNS information series 33 (Lower Hutt: Institute of Geological & Nuclear Sciences, 1995)

10. A.B. Christie & R.G. Barker, *Mineral Resource Assessment of the Northland Region, New Zealand*, GNS Science Report 2007/06 (Lower Hutt: GNS Science, 2007).

11. Simon Hartley, *ODT*, 13 November 2015.

12. Simon Hartley, *ODT*, 18 June 2018.

13. C. Fairburn, 'Mining black pearl', *Quarrying and Mining Magazine*, 13 August 2018: https://quarryingandminingmag.co.nz/mining-black-pearl/

14. Simon Hartley, *ODT*, 28 May 2018; 20 April 2019.

15. Michael Stedman, *ODT*, 24 May 2019; Chris Morris, *ODT*, 24 June 2019.

16. *ODT*, 20 April 2019.

17. Ibid.

18. Rod Oram, *Newsroom*, 24 May 2019.

19. Farah Hancock, *Newsroom*, 17 May 2019.

20. Tim Miller, *ODT,* 14 June 2019.

21. Chris Morris, *ODT*, 5 November 2019.

22. Elena McPhee, *ODT*, 1 August 2019.

CHAPTER 3

1. F.W. Hutton & G.H.F. Ulrich, *Report on the Geology and Gold Fields of Otago* (Dunedin: Provincial Council Otago, 1875).

2. Tréguer, P., Bowler, C., Moriceau, B. et al., 'Influence of diatom diversity on the ocean biological carbon pump', *Nature Geosci 11*, 2018, pp. 27–37: https://doi.org/10.1038/s41561-017-0028-x

3. J.K. Lindqvist & D.E. Lee, 'High-frequency paleoclimate signals from Foulden Maar, Waipiata Volcanic Field, southern New Zealand: An early Miocene varved lacustrine diatomite deposit', *Sedimentary Geology 222*, 2009, pp. 98–110; M.A. Harper, B. Van de Vijver, U. Kaulfuss & D.E. Lee, 'Resolving the confusion between two fossil freshwater diatoms from Otago, New Zealand: *Encyonema jordanii* and *Encyonema jordaniforme* (Cymbellaceae, Bacillariophyta)', *Phytotaxa 394*, 2019, pp. 231–43.

4. Hutton & Ulrich, *Report on the Geology and Gold Fields of Otago*.

5. Harper, Van de Vijver et al., 'Resolving the confusion'.

6. Ibid.

7. Ibid.

8. U. Kaulfuss, 'Geology and paleontology of Foulden Maar, Otago, New Zealand', PhD thesis, University of Otago, Dunedin, 2013.

9. Lindqvist & Lee, 'High-frequency paleoclimate signals from Foulden Maar'.

10. R. Manconi & R. Pronzato, 'Global diversity of sponges (Porifera: Spongillina) in freshwater', *Hydrobiologia 595*, 2008, pp. 27–33.

11. M. Kelly, A.R. Edwards, M.R. Wilkinson et al., 'Phylum Porifera sponges', in *The New Zealand Inventory of Biodiversity Vol. 1: Kingdom Animalia: Radiata, Lophotrochozoa, and Deuterostomia*, D.P. Gordon (ed.) (Christchurch: Canterbury University Press, 2009), pp. 23–46.

12. C. Travis, 'Geology of the Slip Hill area east of Middlemarch', MSc thesis, University of Otago, Dunedin, 1965.

13. R.M. McDowall, *The Reed Field Guide to New Zealand Freshwater Fishes* (Auckland: Reed Publishing, 2000).

14. C. Darwin, *The Origin of Species by Means of Natural Selection*, 6th (reprint) edn (London: J.M. Dent & Sons, 1872).

15. U. Kaulfuss, D.E. Lee, J.H. Robinson et al., 'A review of *Galaxias* (Galaxiidae) fossils from the Southern Hemisphere', *Diversity 12*, 2020, p. 208.

16. D.E. Lee, R.M. McDowall & J.K. Lindqvist, '*Galaxias* fossils from Miocene lake deposits, Otago, New Zealand: The earliest records of the Southern Hemisphere family Galaxiidae (Teleostei)', *JRSNZ 37*, 2007, pp. 109–30; W. Schwarzhans, R.P. Scofield, A.J.D. Tennyson et al., 'Fish remains, mostly otoliths, from the non-marine early Miocene of Otago, New Zealand', *Acta Palaeontologica Polonica 57*, 2011, pp. 319–50.

17. Lee, McDowall & Lindqvist, '*Galaxias* fossils from Miocene lake deposits, Otago, New Zealand'.

18. D.E. Lee, U. Kaulfuss, J.G. Conran et al., 'Biodiversity and palaeoecology of Foulden Maar: An early Miocene *Konservat-Lagerstätte* deposit in southern New Zealand', *Alcheringa 40*, 2016, pp. 525–41; Kaulfuss, Lee, Robinson et al., 'A review of *Galaxias* (Galaxiidae) fossils from the Southern Hemisphere'.

19. McDowall, *The Reed Field Guide to New Zealand Freshwater Fishes*.

20. M. Wuttke, 'Conservation – dissolution – transformation: On the behaviour of biogenic materials during fossilization', in *Messel: An insight into the history of life and of the earth*, S. Schaal & W. Ziegler (eds) (Oxford: Oxford University Press, 1992), pp. 265–75.

21. Lindqvist & Lee, 'High-frequency paleoclimate signals from Foulden Maar'.

22. Kaulfuss, 'Geology and paleontology of Foulden Maar, Otago, New Zealand'.

23. D.J. Jellyman, 'Management and fisheries of Australasian eels (*Anguilla australis, Anguilla dieffenbachii, Anguilla reinhardtii*)', in *Biology and Ecology of Anguillid Eels*, T. Arai (ed.) (London: CRC Press: 2016), pp. 274–90.

24. Ibid.

CHAPTER 4

1. D.C. Mildenhall, E.M. Kennedy, D.E. Lee et al., 'Palynology of the early Miocene Foulden Maar, Otago, New Zealand: Diversity following destruction', *RPP 204*, 2014a, pp. 27–42.

2. Ibid.

3. J.M. Bannister, D.E. Lee & J.G. Conran, 'Lauraceae from rainforest surrounding an early Miocene maar lake, Otago, southern New Zealand', *RPP 178*, 2012, pp. 13–34.

4. M. Pole, 'Plant macrofossils from the Foulden Hills Diatomite (Miocene), Central Otago, New Zealand', *JRSNZ 26*, 1996, pp. 1–39; D.E. Lee, J.G. Conran et al., 'New Zealand Eocene, Oligocene and Miocene macrofossil and pollen records and modern plant distributions in the Southern Hemisphere', *BR 78*, 2012, pp. 235–60; D.E. Lee, U. Kaulfuss et al., 'Biodiversity and palaeoecology of Foulden Maar: An early Miocene *Konservat-Lagerstätte* deposit in southern New Zealand', *Alcheringa 40*, 2016, pp. 525–41.

5. D.E. Lee, J.M. Bannister et al., 'A fossil *Fuchsia* (Onagraceae) flower and an anther mass with *in situ* pollen from the early Miocene of New Zealand', *AmJB 100*, 2013, pp. 2052–65.

6. U. Kaulfuss, D.E. Lee et al., 'Geology and palaeontology of the Hindon Maar Complex: A Miocene terrestrial fossil *Lagerstätte* in southern New Zealand', *PPP 500*, 2018, pp. 52–68; M. Pole, 'Early Miocene flora of the Manuherikia Group, New Zealand. 8. *Nothofagus*', *JRSNZ 23*, 1993c, pp. 329–44; M. Pole, 'Deciduous *Nothofagus* leaves from the Miocene of Cornish Head, New Zealand', *Alcheringa 18*, 1994a, pp. 79–83.

7. Pole, 'Plant macrofossils from the Foulden Hills Diatomite'; Lee, Conran et al., 'New Zealand Eocene, Oligocene and Miocene macrofossil and pollen records; Lee, Kaulfuss et al., 'Biodiversity and palaeoecology of Foulden Maar'.

8. Lee, Kaulfuss, Conran et al., 'Biodiversity and palaeoecology of Foulden Maar'.

9. I. Breitwieser, P.J. Brownsey et al., *Flora of New Zealand Online*, 2010–2021: www.nzflora.info

10. E. Cieraad & D.E. Lee, 'The New Zealand fossil record of ferns for the past 85 million years', *NZJB 44*, 2006, pp. 143–70 and references therein.

11. A.M. Homes, E. Cieraad, D.E. Lee et al., 'A diverse fern flora including macrofossils with *in situ* spores from the Late Eocene of southern New Zealand', *RPP 220*, 2015, pp. 16–28; A.M. Holden, 'Studies in New Zealand Oligocene and Miocene plant macrofossils', PhD thesis, Victoria University, Wellington, 1983; J.G. Conran, J.A. Jackson et al., '*Gleichenia*-like *Korallipteris alineae* sp. nov. macrofossils (Polypodiophyta) from the Miocene Landslip Hill silcrete, New Zealand', *NZJB 55*, 2017, pp. 258–75; M. Pole, 'Early Miocene flora of the Manuherikia Group, New Zealand. 1. Ferns', *JRSNZ 22*, 1992a, pp. 279–86.

12. J.G. Conran, U. Kaulfuss, J.M. Bannister et al., '*Davallia* (Polypodiales: Davalliaceae) macrofossils from early Miocene Otago (New Zealand) with *in situ* spores', *RPP 162*, 2010, pp. 84–94.

13. C. Tsutsumi & M. Kato, 'Evolution of epiphytes in Davalliaceae and related ferns', *BJLS 151*, 2006, pp. 495–510.

14. M.J. von Konrat, J.E. Braggins & P.J. de Lange, '*Davallia* (Pteridophyta) in New Zealand, including description of a new subspecies of *D. tasmanii*', *NZJB 37*, 1999, pp. 579–93.

15. W.L. Testo, A.R. Field, E.B. Sessa & M. Sundue, 'Phylogenetic and morphological analyses support the resurrection of *Dendroconche* and the recognition of two new genera in Polypodiaceae subfamily Microsoroideae', *Systematic Botany* 44, 2019, pp. 737–52; L.R. Perrie, A.R. Field, D.J. Ohlsen & P.J. Brownsey, 'Expansion of the fern genus *Lecanopteris* to encompass some species previously included in *Microsorum* and *Colysis* (Polypodiaceae)', *Blumea* 66, 2021, pp. 242–48.

16. U. Kaulfuss, J.G. Conran, J.M. Bannister et al., 'A new Miocene fern (*Palaeosorum*: Polypodiaceae) from New Zealand bearing in situ spores of *Polypodiisporites*', *NZJB 57*, 2019, pp. 2–17.

17. J. Dawson & R. Lucas, *New Zealand's Native Trees* (Nelson: Craig Potton Publishing, 2011); G.J. Jordan, R.J. Carpenter, J.M. Bannister, D.E Lee, D.C. Mildenhall & R.S. Hill, 'High conifer diversity in Oligo–Miocene New Zealand', *Australian Systematic Botany 24*, 2011, pp. 121–36.

18. For example, M. Pole, 'The New Zealand flora – entirely long-distance dispersal?', *JB 21*, 1994b, pp. 625–35; E. Biffin, R.S. Hill & A.J. Lowe, 'Did kauri (*Agathis*: Araucariaceae) really survive the Oligocene drowning of New Zealand?', *SB 59*, 2010, pp. 594–602 and references.

19. For example, M. Pole, 'Early Miocene flora of the Manuherikia Group, New Zealand. 2. Conifers', *JRSNZ 22*, 1992b, pp. 287–302; M. Pole, 'Miocene conifers from the Manuherikia Group, New Zealand', *JRSNZ 27*, 1997, pp. 355–70; M. Pole, 'Conifer and cycad distribution in the Miocene of southern New Zealand', *AuJB 55*, 2007a, pp. 143–64; M. Pole, 'Plant macrofossils', in *New Zealand Inventory of Biodiversity vol. 3: Kingdoms Bacteria, Protozoa, Chromista, Plantae, Fungi*, D.P. Gordon (ed.) (Christchurch: Canterbury University Press, 2012), pp. 460–75; G.J. Jordan, R.J. Carpenter, J.M. Bannister et al., 'High conifer diversity in Oligo–Miocene New Zealand', *ASB 24*, 2011, pp. 121–36.

20. Mildenhall, Kennedy, Lee et al., 'Palynology of the early Miocene Foulden Maar'.

21. M. Pole, 'Miocene broad-leaved *Podocarpus* from Foulden Hills, New Zealand', *Alcheringa 17*, 1993d, pp. 173–77.

22. M. Pole, 'Miocene conifers from the Manuherikia Group, New Zealand.

23. J.G. Conran, J.M. Bannister, D.E. Lee et al., 'An update of monocot macrofossil data from New Zealand and Australia', *BJLS 178*, 2015, pp. 394–420; J.G. Conran, D.C. Mildenhall, J.I. Raine et al., 'The monocot fossil pollen record of New Zealand and its implications for palaeoclimates and environments', *BJLS 178*, 2015, pp. 421–40 and references therein.

24. Mildenhall, Kennedy, Lee et al., 'Palynology of the early Miocene Foulden Maar, Otago, New Zealand'.

25. J.G. Conran, J.M. Bannister, D.C. Mildenhall et al., 'Leaf fossils of *Luzuriaga* and a monocot flower with in situ pollen of *Liliacidites contortus* Mildenh. & Bannister sp. nov. (Alstroemeriaceae) from the early Miocene', *AmJB 101*, 2014, pp. 141–55.

26. E. Maciunas, J.G. Conran, J.M. Bannister et al., 'Miocene *Astelia* (Asparagales: Asteliaceae) macrofossils from southern New Zealand', *AuSB 24*, 2011, pp. 19–31.

27. J.L. Birch, S.C. Keeley & C.W. Morden, 'Molecular phylogeny and dating of Asteliaceae (Asparagales): *Astelia s.l.* evolution provides insight into the Oligocene history of New Zealand', *MPE 65*, 2012, pp. 102–15.

28. M.J. Thorsen, K.J.M. Dickinson & P.J. Seddon, 'Seed dispersal systems in the New Zealand flora', *PPEES 11*, 2009, pp. 285–309; D. Kelly, J.J. Ladley, A.W. Robertson et al., 'Mutualisms with the wreckage of an avifauna: The status of bird pollination and fruit dispersal in New Zealand', *NZJE 34*, 2010, pp. 66–85.

29. J.G. Conran, '*Paracordyline kerguelensis*, an Oligocene monocotyledon macrofossil from the Kerguélen Islands', *Alcheringa 21*, 1997, pp. 129–40; J.G. Conran & D.C. Christophel, '*Paracordyline aureonemoralis*

(Lomandraceae): An Eocene monocotyledon from South Australia', *Alcheringa 22*, 1998, pp. 351–59.

30. J. Mehl, '*Eoorchis miocaenica* nov. gen., nov. sp. aus dem Ober-Miozän von Öhningen, der bisher älteste fossile Orchideen-Fund', *Berichte aus den Arbeitskreisen Heimische Orchideen (Hanau) 12*, 1984, pp. 9–21.

31. S.R. Ramírez, B. Gravendeel, R.B. Singer et al., 'Dating the origin of the Orchidaceae from a fossil orchid with its pollinator', *Nature 448*, 2007, pp. 1042–45.

32. G.O. Poinar Jr, 'Beetles with orchid pollinaria in Dominican and Mexican amber', *American Entomologist 62*, 2016a, pp. 172–77; G.O. Poinar Jr, 'Orchid pollinaria (Orchidaceae) attached to stingless bees (Hymenoptera: Apidae) in Dominican amber', *Neues Jahrbuch für Geologie und Paläontologie, Abhandlungen 279*, 2016b, pp. 287–93; G.O. Poinar Jr & F.N. Rasmussen, 'Orchids from the past, with a new species in Baltic amber', *BJLS 183*, 2017, pp. 327–33.

33. A.B. Massalongo, 'Vorläufige Nachricht über die neueren paläontologischen Entdeckungen am Monte Bolca', *Neues Jahrbuch für Mineralogie, Geognosie, Geologie und Petrefakten-kunde 1857*, 1857, pp. 775–78.

34. J.G. Conran, J.M. Bannister & D.E. Lee, 'Earliest orchid macrofossils: Early Miocene *Dendrobium* and *Earina* (Orchidaceae: Epidendroideae) from New Zealand', *AmJB 96*, 2009, pp. 466–74.

35. M.W. Chase, M.J.M. Christenhusz & T. Mirenda, 'Orchid evolution', in *The Book of Orchids: A life-size guide to six hundred species from around the world*, M.W. Chase, M.J.M. Christenhusz & T. Mirenda (eds) (Chicago: University of Chicago Press, 2017), pp. 12–13.

36. NZ Native Orchid Group, *New Zealand Native Orchids*, 2021: www.nativeorchids.co.nz

37. Lee, Kaulfuss, Conran et al., 'Biodiversity and palaeoecology of Foulden Maar'.

38. B.W. Macmillan, 'Biological flora of New Zealand. 7. *Ripogonum scandens* J.R. et G. Forst. (Smilacaceae), Supplejack, Kareao', *NZJB 10*, 1972, pp. 641–72.

39. J.R. Forster & G. Forster, *Characteres generum plantarum quas in itinere ad insulas Maris Australis, collegerunt, descripserunt, delinearunt, annis MDCCLXXII–MDCCLXXV / Joannes Reinoldus Forster et Georgius Forster*, 2nd edn (Londini: Prostant apud B. White, T. Cadell & P. Elmsly, 1776).

40. I.A. Kerr, J.G. Conran, D.E. Lee & M. Waycott, 'Phylogeny, fossil history and biogeography of Ripogonaceae', in *Abstracts, Geosciences 2016, Wanaka, Geoscience Society of New Zealand annual conference, 28 Nov–1 Dec 2016. GSNZ Miscellaneous Publication 145A*, C. Riesselmann & A. Roben (eds) (Wanaka: GSNZ, 2016), p. 103.

41. J.G. Conran, R.J. Carpenter & G.J. Jordan, 'Early Eocene *Ripogonum* (Liliales: Ripogonaceae) leaf macrofossils from southern Australia', *ASB 22*, 2009, pp. 219–28; R.J. Carpenter, P. Wilf, J.G. Conran & N.R. Cúneo, 'A Paleogene trans-Antarctic distribution for *Ripogonum* (Ripogonaceae: Liliales)?', *Palaeontologia Electronica 17*, 2014, pp. 39A, 1–9; J.G. Conran, E.M. Kennedy & J.M. Bannister, 'Early Eocene Ripogonaceae leaf macrofossils from New Zealand', *ASB 31*, 2018, pp. 8–15.

42. Pole, 'Plant macrofossils from the Foulden Hills Diatomite'; Lee, Conran, Lindqvist et al., 'New Zealand Eocene, Oligocene and Miocene macrofossil and pollen records and modern plant distributions in the Southern Hemisphere'; Lee, Kaulfuss, Conran et al., 'Biodiversity and palaeoecology of Foulden Maar'.

43. J.I. Raine, D.C. Mildenhall & E.M. Kennedy, *New Zealand Fossil Spores and Pollen: An illustrated catalogue*, 4th edn (GNS Science Miscellaneous Series No. 4), 2011: www.gns.cri.nz/what/earthhist/fossils/spore_pollen/catalog/index.htm, 853 html pages; issued also in CD version; Mildenhall, Kennedy, Lee et al., 'Palynology of the early Miocene Foulden Maar'; Conran, Mildenhall, Raine et al., 'The monocot fossil pollen record of New Zealand'.

44. W.R.B. Oliver, 'The flora of the Waipaoa Series (later Pliocene) of New Zealand', *Transactions of the New Zealand Institute 59*, 1928, pp. 287–303; M. Pole, 'Monocot macrofossils from the Miocene of southern New Zealand', *PE 10*, 2007b, pp. 3.15A, 1–21.

45. Cieraad & Lee, 'The New Zealand fossil record of ferns for the past 85 million years'.

CHAPTER 5

1. B.E. Balme, 'Fossil *in situ* spores and pollen grains: An annotated catalogue', *RPP 87*, 1995, pp. 81–323.

2. E.M. Friis, P.R. Crane & K.R. Pedersen, *Early Flowers and Angiosperm Evolution* (Cambridge: Cambridge University Press, 2011).

3. M. Pole, B. Douglas & G. Mason, 'The terrestrial Miocene biota of southern New Zealand', *JRSNZ 33*, 2003, pp. 415–26; E.M. Kennedy, J.D. Lovis & I.L. Daniel, 'Discovery of a Cretaceous angiosperm reproductive structure from New Zealand', *NZJGG 46*, 2003, pp. 519–22.

4. J.M. Bannister, D.E. Lee & J.I. Raine, 'Morphology and palaeoenvironmental context of *Fouldenia staminosa*, a fossil flower with associated pollen from the early Miocene of Otago, New Zealand', *NZJB 43*, 2005, pp. 515–25.

5. K. Faegri & L. van der Pijl, *The Principles of Pollination Ecology*, 3rd rev. edn (Oxford: Pergamon, 1979).

6. J.G. Conran, U. Kaulfuss, J.M. Bannister et al., 'An *Akania* (Akaniaceae) inflorescence with associated pollen from the early Miocene of New Zealand', *AmJB 106*, 2019, pp. 1–11. *Akania gibsonorum* was named in honour of the Gibson family.

7. E.J. Romero & L.J. Hickey, 'A fossil leaf of Akaniaceae from Paleocene beds in Argentina', *Bulletin of the Torrey Botanical Club 103*, 1976, pp. 126–31; M.A. Gandolfo, M.C. Dibbern & E.J. Romero, '*Akania patagonica n. sp.* and additional material on *Akania americana* Romero & Hickey (Akaniaceae), from Paleocene sediments of Patagonia', *Bulletin of the Torrey Botanical Club 115*, 1988, pp. 83–88; M. Brea, A.F. Zucol, M.S. Bargo et al., 'First Miocene record of Akaniaceae in Patagonia (Argentina): A fossil wood from the early Miocene Santa Cruz formation and its palaeobiogeographical implications', *BJLS 183*, 2017, pp. 334–47.

8. L.D. Shepherd, P.J. de Lange et al., 'A biological and ecological review of the endemic New Zealand genus *Alseuosmia* (toropapa; Alseuosmiaceae)', *NZJB 58*, 2020, pp. 2–18.

9. J.G. Conran, W.G. Lee, D.E. Lee et al., 'Reproductive niche conservatism in the isolated New Zealand flora over 23 million years', *Biology Letters 10*, 2014, p. 20140647.

10. J. Dawson & R. Lucas, *New Zealand's Native Trees* (Nelson: Craig Potton Publishing, 2011).

11. Ibid.

12. J.G. Conran, J.M. Bannister & D.E. Lee, 'Fruits and leaves with cuticle of *Laurelia otagoensis sp. nov.* (Atherospermataceae) from the early Miocene of Otago (New Zealand)', *Alcheringa 37*, 2013, pp. 496–509.

13. Conran, Lee, Lee et al., 'Reproductive niche conservatism in the isolated New Zealand flora over 23 million years'.

14. J.A. Hunter, 'A further note on *Tecomanthe speciosa* W.R.B. Oliver (Bignoniaceae)', *Records of the Auckland Institute and Museum 6*, 1967, pp. 169–70.

15. W. Cooper & W.T. Cooper, *Fruits of the Australian Tropical Rainforest* (Melbourne: Nokomis Editions, 2004).

16. J.C. Bradford, H.C.F. Hopkins & R.W. Barnes, 'Cunoniaceae', in *The Families and Genera of Vascular Plants vol. 6. Flowering plants. Dicotyledons: Celastrales, Oxalidales, Rosales, Cornales, Ericales*, K. Kubitzki (ed.) (Berlin: Springer Verlag, 2004), pp. 91–111.

17. Dawson & Lucas, *New Zealand's Native Trees*.

18. P. Wardle & A.H. MacRae, 'Biological flora of New Zealand. 1. *Weinmannia racemosa*', *NZJB 4*, 1966, pp. 114–31.

19. E.M. Friis, P.R. Crane & K.R. Pedersen, *Early Flowers and Angiosperm Evolution* (Cambridge: Cambridge University Press, 2011).

20. M.J.E. Coode, 'Elaeocarpaceae', in *The Families and Genera of Vascular Plants vol. 6. Flowering plants. Dicotyledons: Celastrales, Oxalidales, Rosales, Cornales, Ericales*, K. Kubitzki (ed.) (Berlin: Springer Verlag, 2004), pp. 135–44.

21. D.J. Mabberley, *Mabberley's Plant-book: A portable dictionary of plants, their classification and uses*, 4th edn (Cambridge: Cambridge University Press, 2017); M.E. Dettmann & H.T. Clifford, 'The fossil record of *Elaeocarpus* L. fruits', *Memoirs of the Queensland Museum 46*, 2000, pp. 461–97; A.C. Rozefelds & D.C. Christophel, 'Cenozoic *Elaeocarpus* (Elaeocarpaceae) fruits from Australia', *Alcheringa 26*, 2002, pp. 261–74.

22. M. Pole, 'Plant macrofossils from the Foulden Hills Diatomite (Miocene), Central Otago, New Zealand', *JRSNZ 26*, 1996, pp. 1–39; A.C. Harris, J.M. Bannister & D.E. Lee, 'Fossil scale insects (Hemiptera, Coccoidea, Diaspididae) in life position on an angiosperm leaf from an early Miocene lake deposit, Otago, New Zealand', *JRSNZ 37*, 2007, pp. 1–13.

23. Pole, 'Plant macrofossils from the Foulden Hills Diatomite'; D.E. Lee, J.M. Bannister et al., 'Euphorbiaceae: Acalyphoideae fossils from early Miocene New Zealand: *Mallotus–Macaranga* leaves, fruits, and inflorescence with *in situ Nyssapollenites endobalteus* pollen', *RPP 163*, 2010, pp. 127–38.

24. M. Nucete, J.H.A van Konijnenburg-van Cittert & P.C. van Welzen, 'Fossils and palaeontological distributions of *Macaranga* and *Mallotus* (Euphorbiaceae)', *PPP 353–55*, 2012, pp. 104–15.

25. E. Yamasaki & S. Sakai, 'Wind and insect pollination (ambophily) of *Mallotus* spp. (Euphorbiaceae) in tropical and temperate forests', *AuJB 61*, 2013, pp. 60–66.

26. J.M. Bannister, D.E. Lee & J.G. Conran, 'Lauraceae from rainforest surrounding an early Miocene maar lake, Otago, southern New Zealand', *RPP 178*, 2012, pp. 13–34.

27. M.K. MacPhail, 'Fossil and modern *Beilschmiedia* (Lauraceae) pollen in New Zealand', *NZJB 18*, 1980, pp. 453–57.

28. Bannister, Lee & Conran, 'Lauraceae from rainforest surrounding an early Miocene maar lake'; D.C. Mildenhall, E.M. Kennedy, D.E. Lee et al., 'Palynology of the early Miocene Foulden Maar, Otago, New Zealand: Diversity following destruction', *RPP 204*, 2014, pp. 27–42.

29. T.H. Worthy, S.J. Hand, J.P. Worthy et al., 'A large fruit pigeon (Columbidae) from the early Miocene of New Zealand', *The Auk 126*, 2009, pp. 649–56.

30. Ibid.

31. Bannister, Lee, & Conran, 'Lauraceae from rainforest surrounding an early Miocene maar lake.

32. Worthy et al., 'A large fruit pigeon'.

33. Conran, Lee, Lee et al., 'Reproductive niche conservatism in the isolated New Zealand flora over 23 million years'.

34. D. Kelly, J.J. Ladley & A.W. Robertson, 'Is dispersal easier than pollination? Two tests in New Zealand Loranthaceae', *NZJB 42*, 2004, pp. 89–103.

35. D.J. Mabberley, 'Meliaceae', in *The Families and Genera of Vascular Plants vol. 10. Flowering Plants. Eudicots: Sapindales, Cucurbitales, Myrtaceae*, K. Kubitzki (ed.) (Berlin: Springer Verlag, 2011), pp. 185–211.

36. Dawson & Lucas, *New Zealand's Native Trees*; D. Kelly, J.J. Ladley, A.W. Robertson et al., 'Mutualisms with the wreckage of an avifauna: The status of bird pollination and fruit dispersal in New Zealand', *NZJE 34*, 2010, pp. 66–85.

37. E.J.M. Koenen, J.J. Clarkson et al., 'Recently evolved diversity and convergent radiations of rainforest mahoganies (Meliaceae) shed new light on the origins of rainforest hyperdiversity', *New Phytologist 207*, 2015, pp. 327–39; M. Sun, R. Naeem, J.-X. Su et al., 'Phylogeny of the Rosidae: A dense taxon sampling analysis', *Journal of Systematics and Evolution 54*, 2016, pp. 363–91.

38. Conran, Lee, Lee et al., 'Reproductive niche conservatism in the isolated New Zealand flora over 23 million years'; Kelly, Ladley & Robertson, 2004.

39. M.J.M. Christenhusz & J.W. Byng, 'The number of known plants species in the world and its annual increase', 2016, pp. 201–17.

40. P.J.A. Kessler, 'Menispermaceae', in *The Families and Genera of Vascular Plants vol. 2. Flowering plants. Dicotyledons: Magnoliid, hamamelid and caryophyllid families*, K. Kubitzki, J.G. Rohwer & V. Bittrich (eds), (Berlin: Springer-Verlag, 1993), pp. 402–18.

41. J. Jérémie, 'Étude des Monimiaceae: Révision du genre *Hedycarya*', *Adansonia, séries 2 18*, 1978, pp. 25–53.

42. Dawson & Lucas, *New Zealand's Native Trees*.

43. S.A. Norton, 'Thrips pollination in the lowland forest of New Zealand', *NZJE 7*, 1984, pp. 157–64.

44. Kelly, Ladley & Robertson, 'Is dispersal easier than pollination?'

45. S.S. Renner, J.S. Strijk et al., 'Biogeography of the Monimiaceae (Laurales): A role for East Gondwana and long-distance dispersal, but not West Gondwana', *JB 37*, 2010, pp. 1227–38.

46. M.J.M. Christenhusz & J.W. Byng, 'The number of known plants species in the world and its annual increase', *Phytotaxa 261*, 2016, pp. 201–17.

47. M. Pole, 'Early Miocene flora of the Manuherikia Group, New Zealand. 7. Myrtaceae, including *Eucalyptus*', *Journal of the Royal Society of New Zealand 23*, 1993b, pp. 313–28;

M. Pole, 'Dispersed leaf cuticle from the early Miocene of southern New Zealand', *PE 11*, 2008, pp. 15A, 1–117; D.E. Lee, J.G. Conran, J.K. Lindqvist et al., 'New Zealand Eocene, Oligocene and Miocene macrofossil and pollen records and modern plant distributions in the Southern Hemisphere', *Botanical Review 78*, 2012, pp. 235–60.

48. M. Pole, J. Dawson & T. Denton, 'Fossil Myrtaceae from the early Miocene of southern New Zealand', *AuJB 56*, 2008, pp. 67–81.

49. Y. Pillon, E. Lucas, J.B. Johansen et al., 'An expanded *Metrosideros* (Myrtaceae) to include *Carpolepis* and *Tepualia* based on nuclear genes', *SB 40*, 2015, pp. 782–90.

50. P.E. Berry, W.J. Hahn, K.J. Sytsma et al., 'Phylogenetic relationships and biogeography of *Fuchsia* (Onagraceae) based on noncoding nuclear and chloroplast DNA data', *AmJB 91*, 2004, pp. 601–14.

51. Kelly, Ladley & Robertson, 'Is dispersal easier than pollination?'

52. Ibid.

53. E.J. Godley & P.E. Berry, 'The biology and systematics of *Fuchsia* in the South Pacific', *Annals of the Missouri Botanical Garden 82*, 1995, pp. 473–516.

54. Angiosperm Phylogeny Group, 'An update of the Angiosperm Phylogeny Group classification for the orders and families of flowering plants: APG III', *BJLS 161*, 2009, pp. 105–21.

55. A.I. Ruiz, M.E. Guantay & G.I. Ponessa, 'Leaf morphology, anatomy and foliar architecture of *Myrsine laetevirens* (Myrsinaceae)', *Lilloa 49*, 2012, pp. 59–67.

56. Pole, 'Plant macrofossils from the Foulden Hills Diatomite'; Pole, 'Dispersed leaf cuticle from the early Miocene of southern New Zealand'; M. Pole & P.R. Moore, 'A late Miocene leaf assemblage from Coromandel Peninsula, New Zealand, and its climatic implications', *Alcheringa 35*, 2010, pp. 103–21.

57. B.A. Meylan & B.G. Butterfield, 'The structure of New Zealand woods', *New Zealand Department of Science and Industry Research Bulletin 222*, 1978, pp. 1–250; F. Lens, S. Jansen, P. Caris et al., 'Comparative wood anatomy of the Primuloid Clade (Ericales s.l)', *SB 30*, 2005, pp. 163–83.

58. D.E. Lee, U. Kaulfuss, J.G. Conran et al., 'Biodiversity and palaeoecology of Foulden Maar: An early Miocene Konservat-Lagerstätte deposit in southern New Zealand', *Alcheringa 40*, 2016, pp. 525–41.

59. M.J.M. Christenhusz, M.F. Fay & M.W. Chase, *Plants of the World: An illustrated encyclopedia of vascular plants* (Richmond, UK: RBG Kew and Chicago: University of Chicago Press, 2017).

60. Dawson & Lucas, *New Zealand's Native Trees*.

61. M. Pole, 'The Proteaceae record in New Zealand', *ASB 11*, 1998, pp. 343–72; Pole, 'Dispersed leaf cuticle from the early Miocene of southern New Zealand'; H. Sauquet, P.H. Weston, C.L. Anderson et al., 'Contrasted patterns of hyperdiversification in Mediterranean hotspots', *Proceedings of the National Academy of Sciences 106*, 2009, pp. 221–25.

62. For more detailed discussion and additional references see R.J. Carpenter, J.M. Bannister, D.E. Lee & G.J. Jordan, 'Proteaceae leaf fossils from the Oligo-Miocene of New Zealand: New species and evidence of biome and trait conservatism', *ASB 25*, 2012, pp. 375–89.

63. Pole, 'The Proteaceae record in New Zealand'; for more detailed discussion of species of Proteaceae and additional references see Carpenter, Bannister et al., 'Proteaceae leaf fossils from the Oligo-Miocene of New Zealand', 2012.

64. P.K. Buchanan, R.E. Beever, T.R. Glare et al., 'Kingdom fungi: Introduction', in *New Zealand Inventory of Biodiversity vol. 3: Kingdoms Bacteria, Protozoa, Chromista, Plantae, Fungi*, D.P. Gordon (ed.) (Christchurch: Canterbury University Press, 2012), pp. 499–515; C. Shirley, *Forest Fungi Quick Reference Guide, version 5.01* (published online by the author, 2020): http://hiddenforest.co.nz/fungi/

65. J.M. Bannister, J.G. Conran & D.E. Lee, 'Life on the phylloplane: Eocene epiphyllous fungi from Pikopiko Fossil Forest, Southland, New Zealand', *NZJB 54*, 2016, pp. 412–32; J.G. Conran, J.M. Bannister, T. Reichgelt & D.E. Lee, 'Epiphyllous fungi and leaf physiognomy indicate an ever-wet humid mesothermal (subtropical) climate in the late Eocene of southern New Zealand', *PPP 452*, 2016, pp. 1–10.

66. T. Reichgelt, E.M. Kennedy, J.G. Conran et al., 'The presence of moisture deficits in Miocene New Zealand', *GPC 172*, 2019, pp. 268–77.

CHAPTER 6

1. R.P. Macfarlane P.A. Maddison, I.G. Andrew et al., 'Phylum Arthropoda subphylum Hexapoda: Protura, springtails, Diplura, and insects', in *New Zealand Inventory of Biodiversity vol. 2: Kingdom Animalia: Chaetognatha, Ecdysozoa, ichnofossils*, D.P. Gordon (ed.) (Christchurch: Canterbury University Press, 2011), pp. 233–467; P.J. Sirvid, Z.-Q. Zhang, M.S. Harvey et al., 'Phylum Arthropoda Chelicerata: Horseshoe crabs, arachnids, sea spiders', in *New Zealand Inventory of Biodiversity vol. 2: Kingdom Animalia: Chaetognatha, Ecdysozoa, ichnofossils*, D.P. Gordon (ed.) (Christchurch: Canterbury University Press, 2011), pp. 50–89.

2. U. Kaulfuss, 'Geology and paleontology of Foulden Maar, Otago, New Zealand', PhD thesis, University of Otago, Dunedin, 2013.

3. Sirvid, Zhang, Harvey et al., 'Phylum Arthropoda Chelicerata'.

4. Ibid.; P.A. Selden & U. Kaulfuss, 'Fossil arachnids from the earliest Miocene Foulden Maar Fossil-Lagerstätte, New Zealand', *Alcheringa 43*, 2018, pp. 165–69.

5. Macfarlane, Maddison, Andrew et al., 'Phylum Arthropoda subphylum Hexapoda'.

6. J.I. Sutherland, 'Miocene petrified wood and associated borings and termite faecal pellets from Hukatere Peninsula, Kaipara Harbour, North Auckland, New Zealand', *JRSNZ 33*, 2003, pp. 395–414.

7. U. Kaulfuss, A.C. Harris & D.E. Lee, 'A new fossil termite (Isoptera, Stolotermitidae, *Stolotermes*) from the early Miocene of Otago, New Zealand', *Acta Geologica Sinica 84*, 2010, pp. 705–09; M.S. Engel & U. Kaulfuss, 'Diverse, primitive termites (Isoptera: Kalotermitidae, *incertae sedis*) from the early Miocene of New Zealand', *Austral Entomology 56*, 2016, pp. 94–103.

8. Macfarlane, Maddison, Andrew et al., 'Phylum Arthropoda subphylum Hexapoda'.

9. A.R. Schmidt, U. Kaulfuss, J.M. Bannister et al., 'Amber inclusions from New Zealand', *Gondwana Research 56*, 2018, pp. 135–46; U. Kaulfuss, S.D.J. Brown, I.M. Henderson et al., 'First insects from the Manuherikia Group, early Miocene, New Zealand', *JRSNZ 49*, 2018, pp. 494–507; U. Kaulfuss & M. Moulds, 'A new genus and species of tettigarctid cicada from the early Miocene of New Zealand: *Paratettigarcta zealandica* (Hemiptera, Auchenorrhyncha, Tettigarctidae)', *ZooKeys 484*, 2015, pp. 83–94.

10. A.C. Harris, J.M. Bannister & D.E. Lee, 'Fossil scale insects (Hemiptera, Coccoidea, Diaspididae) in life position on an angiosperm leaf from an early Miocene lake deposit, Otago, New Zealand', *JRSNZ 37*, 2007, pp. 1–13.

11. U. Kaulfuss, T. Wappler, E. Heiss & M.-C. Larivière, '*Aneurus* sp. from the early Miocene Foulden Maar, New Zealand: The first Southern Hemisphere record of fossil Aradidae (Insecta: Hemiptera: Heteroptera)', *JRSNZ 41*, 2011, pp. 279–85.

12. M.-C. Larivière & A. Larochelle, 'An overview of flat bug genera (Hemiptera, Aradidae) from New Zealand, with considerations on faunal diversification and affinities', *Denisia 19*, 2006, pp. 181–214.

13. M.-C. Larivière, 'Cixiidae (Insecta: Hemiptera, Auchenorrhyncha)', *Fauna of New Zealand 40*, 1999, pp. 1–93.

14. M.-C. Larivière, M.J. Fletcher & A. Larochelle, 'Auchenorrhyncha (Insecta: Hemiptera): Catalogue', *Fauna of New Zealand 63*, 2010, pp. 1–232.

15. U. Kaulfuss, D.E. Lee et al., 'A diverse fossil terrestrial arthropod fauna from New Zealand: Evidence from the early Miocene Foulden Maar fossil *Lagerstätte*', *Lethaia 48*, 2014, pp. 299–308.

16. G. Kuschel, 'Beetles in a suburban environment: A New Zealand case study. The identity and status of Coleoptera in the natural and modified habitats of Lynfield, Auckland (1974–1989)', *DSIR Plant Protection Report 3*, 1990, pp. 1–119; R.A.B. Leschen, J.F. Lawrence, G. Kuschel et al., 'Coleoptera genera of New Zealand', *New Zealand Entomologist 26*, 2003, pp. 15–28; Macfarlane, Maddison, Andrew et al., 'Phylum Arthropoda subphylum Hexapoda'.

17. U. Kaulfuss, D.E. Lee et al., 'Geology and palaeontology of the Hindon Maar Complex: A Miocene terrestrial fossil *Lagerstätte* in southern New Zealand', *PPP 500*, 2018, pp. 52–68.

18. Kaulfuss, Brown, Henderson et al., 'First insects from the Manuherikia Group, early Miocene, New Zealand'.

19. J. Jenkins Shaw, A. Solodovnikov, M. Bai & U. Kaulfuss, 'An amblyopinine rove beetle (Coleoptera, Staphylinidae, Staphylininae, Amblyopinini) from the earliest Miocene Foulden Maar fossil-Lagerstätte, New Zealand', *Journal of Paleontology 94*, 2020, pp. 1082–88.

20. AntWiki: www.antwiki.org

21. A.A. Forbes, R.K. Bagley, M.A. Beer et al., 'Quantifying the unquantifiable: Why Hymenoptera, not Coleoptera, is the most speciose animal order', *BMC Ecology 18*, 2018, p. 21.

22. Macfarlane, Maddison, Andrew et al., 'Phylum Arthropoda subphylum Hexapoda'.

23. U. Kaulfuss, A.C. Harris, J.G. Conran & D.E. Lee, 'An early Miocene ant (subfam. Ambyloponinae) from Foulden Maar: The first fossil Hymenoptera from New Zealand', *Alcheringa 38*, 2014, pp. 568–74; U. Kaulfuss & G.M. Dlussky, 'Early Miocene Formicidae (Amblyoponinae, Ectatomminae, ?Dolichoderinae, Formicinae, and Ponerinae) from the Foulden Maar Fossil *Lagerstätte*, New Zealand, and their biogeographic relevance', *Journal of Paleontology 89*, 2015, pp. 1043–55.

24. J.C. Morse, P.B. Frandsen, W. Graf & J.A. Thomas, 'Diversity and ecosystem services of Trichoptera', *Insects 10*, 2019, p. 125.

25. Macfarlane, Maddison, Andrew et al., 'Phylum Arthropoda subphylum Hexapoda'.

26. H. de Jong, P. Oosterbroek, J. Gelhaus et al., 'Global diversity of craneflies (Insecta, Diptera: Tipulidea or Tipulidae sensu lato) in freshwater', in *Freshwater Animal Diversity Assessment. Developments in Hydrobiology*, vol. 198, E.V. Balian, C. Lévéque et al. (eds), (Dordrecht: Springer, 2007).

27. Macfarlane, Maddison, Andrew et al., 'Phylum Arthropoda subphylum Hexapoda'.

28. A.C. Harris, 'An Eocene larval insect fossil (Diptera Bibionidae) from north Otago, New Zealand', *JRSNZ 13*, 1983, pp. 93–105.

29. Macfarlane, Maddison, Andrew et al., 'Phylum Arthropoda subphylum Hexapoda'.

30. de Jong, Oosterbroek, Gelhaus et al., 'Global diversity of craneflies (Insecta, Diptera: Tipulidea or Tipulidae sensu lato) in freshwater'.

31. Macfarlane, Maddison, Andrew et al., 'Phylum Arthropoda subphylum Hexapoda'.

32. Ibid.

33. S.B. Waters, 'A Cretaceous dance fly (Diptera: Empididae) from Botswana', *Systematic Entomology 14*, 1989, pp. 233–41.

34. J.G. Conran, W.G. Lee, D.E. Lee et al., 'Reproductive niche conservatism in the isolated New Zealand flora over 23 million years', *Biology Letters 10*, 2014, p. 20140647.

35. F. Grímsson, R. Zetter, C.C. Labandeira et al., 'Taxonomic description of *in situ* bee pollen from the middle Eocene of Germany', *Grana 56*, 2017, pp. 37–70.

36. Kaulfuss, 'Geology and paleontology of Foulden Maar, Otago, New Zealand'.

CHAPTER 7

1. M. Grein, W. Konrad, V. Wilde et al., 'Reconstruction of atmospheric CO_2 during the early middle Eocene by application of a gas exchange model to fossil plants from the Messel Formation, Germany', *PPP 309*, 2011, pp. 383–91; O.K. Lenz, V. Wilde & W. Riegel, 'Short-term fluctuations in vegetation and phytoplankton during the Middle Eocene greenhouse climate: A 640-kyr record from the Messel oil shale (Germany)', *International Journal of Earth Sciences 100*, 2011, pp. 1851–74; O.K. Lenz, V. Wilde & W. Riegel, 'ENSO- and solar-driven sub-Milankovitch cyclicity in the Palaeogene greenhouse world: High-resolution pollen records from Eocene Lake Messel, Germany', *Journal of the Geological Society 174*, 2016, pp. 110–28; O.K Lenz & V. Wilde, 'Changes in Eocene plant diversity and composition of vegetation: The lacustrine archive of Messel (Germany)', *Paleobiology 44*, 2018, pp. 709–35.

2. M.E. Raymo, L.E. Lisiecki & K.H. Nisancioglu, 'Plio-Pleistocene ice volume, Antarctic climate, and the global δ18O record', *Science 313*, 2006, pp. 492–95; J. Jouzel, V. Masson-Delmotte, O. Cattani et al., 'Orbital and millennial

Antarctic climate variability over the past 800,000 years', *Science 317*, 2007, pp. 793–96; D. Lüthi, M. Le Floch, B. Bereiter et al., 'High-resolution carbon dioxide concentration record 650,000–800,000 years before present', *Nature 453*, 2008, pp. 379–82; R.J. Carpenter, G.J. Jordan, M.K. MacPhail & R.S. Hill, 'Near-tropical early Eocene terrestrial temperatures at the Australo-Antarctic margin, western Tasmania', *Geology 40*, 2012, pp. 267–70.

3. B.R.S. Fox, G.S. Wilson & D.E. Lee, 'A unique annually laminated maar lake sediment record shows orbital control of Southern Hemisphere mid-latitude climate across the Oligocene–Miocene boundary', *GSAB 128*, 2016, pp. 609–26.

4. B.R.S. Fox, J. Wartho, G.S. Wilson et al., 'Long-term evolution of an Oligocene/Miocene maar lake from Otago, New Zealand', *GGG 16*, 2015, pp. 59–76; Fox, Wilson & Lee, 'A unique annually laminated maar lake sediment record'.

5. J.A. Wolfe, 'A method of obtaining climatic parameters from leaf assemblages', *Bulletin of the United States Geological Survey 2040*, 1993, pp. 1–71.

6. T. Reichgelt, E.M. Kennedy, D.C. Mildenhall et al., 'Quantitative palaeoclimate estimates for early Miocene southern New Zealand: Evidence from Foulden Maar', *PPP 378*, 2013, pp. 36–44.

7. Carpenter, Jordan, MacPhail et al., 'Near-tropical early Eocene terrestrial temperatures at the Australo-Antarctic margin, western Tasmania'.

8. T. Reichgelt, E.M. Kennedy, D.C. Mildenhall et al., 'Quantitative palaeoclimate estimates for early Miocene southern New Zealand'.

9. Fox, Wilson & Lee, 'A unique annually laminated maar lake sediment record'; B.R.S. Fox, W.J. D'Andrea, G.S. Wilson et al., 'Interaction of polar and tropical influences in the mid-latitudes of the Southern Hemisphere during the Mi-1 deglaciation', *GPC 155*, 2017, pp. 109–20.

10. Data compiled from J.C. Zachos, G.R. Dickens & R.E. Zeebe, 'An early Cenozoic perspective on greenhouse warming and carbon-cycle dynamics', *Nature 451*, 2008, pp. 279–83; J. Hansen, M. Sato, G. Russell & P. Kharecha, 'Climate sensitivity, sea level and atmospheric carbon dioxide', *Philosophical Transactions of the Royal Society A: Mathematical, Physical and Engineering Sciences 371*, 2013. Redrawn by Tammo Reichgelt.

11. G.L. Foster, D.L. Royer & D.J. Lunt, 'Future climate forcing potentially without precedent in the last 420 million years', *Nature Communications 8*, 2017, p. 14845.

12. Y.G. Zhang, M. Pagani, Z. Liu et al., 'A 40-million-year history of atmospheric CO_2', *Philosophical Transactions of the Royal Society A: Mathematical, Physical and Engineering Sciences 371*, 2013.

13. T. Reichgelt, W.J. D'Andrea & B.R.S. Fox, 'Abrupt plant physiological changes in southern New Zealand at the termination of the Mi-1 event reflect shifts in hydroclimate and pCO_2', *Earth and Planetary Science Letters 455*, 2016, pp. 115–24; T. Reichgelt, W.J. D'Andrea, A.C. Valdivia-McCarthy et al., 'Elevated CO_2, increased leaf-level productivity, and water-use efficiency during the early Miocene', *Climate of the Past 16*, 2020, pp. 1509–21.

14. See Reichgelt, D'Andrea et al., 'Elevated CO_2, increased leaf-level productivity, and wateruse efficiency during the early Miocene', *Climate of the Past* 16, 2020, pp. 1509–21.

15. G.S. Wilson, S.F. Pekar, T.R. Naish et al., 'The Oligocene–Miocene boundary – Antarctic climate response to orbital forcing', in *Developments in Earth and Environmental Sciences vol. 8*, F. Florindo & M. Siegert (eds), (Amsterdam: Elsevier, 2008), pp. 369–400; Reichgelt, D'Andrea & Fox, 2016.

16. Reichgelt, D'Andrea et al., 'Elevated CO_2, increased leaf-level productivity, and water-use efficiency during the early Miocene'.

17. Ibid.

18. J.K. Lindqvist & D.E. Lee, 'High-frequency paleoclimate signals from Foulden Maar, Waipiata Volcanic Field, southern New Zealand: An early Miocene varved lacustrine diatomite deposit', *Sedimentary Geology 222*, 2009, pp. 98–110; B.R.S. Fox, 'Climate change at the Oligocene/Miocene boundary'. PhD thesis, University of Otago, Dunedin, 2014.

19. Reichgelt, D'Andrea et al., 'Elevated CO_2, increased leaf-level productivity, and water-use efficiency during the early Miocene'.

CHAPTER 8

1. N. Mortimer & H.J. Campbell, *Zealandia: Our continent revealed* (Auckland: Penguin, 2014); N. Mortimer, H.J. Campbell, A.J. Tulloch et al., 'Zealandia: Earth's hidden continent', *GSA Today 27*, 2017, pp. 27–35.

2. G.W. Gibbs, *Ghosts of Gondwana: The history of life in New Zealand*, rev. edn, (Nelson: Craig Potton Publishing, 2016).

3. D.E. Lee, W.G. Lee & N. Mortimer, 'Where and why have all the flowers gone? Depletion and turnover in the New Zealand Cenozoic angiosperm flora in relation to palaeogeography and climate', *AuJB 49*, 2001, pp. 341–56.

4. D.E. Lee, W.G. Lee, G.J. Jordan & V.D. Barreda, 'The Cenozoic history of New Zealand temperate rainforests: Comparisons with southern Australia and South America', *NZJB 54*, 2016, pp. 100–27.

5. P.J.J. Kamp, K.A. Vincent & M.J.S. Tayler, *Cenozoic Sedimentary and Volcanic Rocks of New Zealand: A reference volume of lithology, age and paleoenvironments with maps (PMAPs) and database* (Hamilton: University of Waikato, 2015).

6. Gibbs, *Ghosts of Gondwana*.

7. e.g. M. Pole, 'The New Zealand flora – entirely long-distance dispersal?', *JB 21*, 1994b, pp. 625–35; S.A. Trewick, A.M. Paterson & H.J. Campbell, 'Hello New Zealand', *JB 34*, 2007, pp. 1–6; H. Campbell, *The Zealandia Drowning Debate: Did New Zealand sink beneath the waves?* (Wellington: Bridget Williams Books, 2013).

8. H. Campbell & G. Hutching, *In Search of Ancient New Zealand* (Auckland: Penguin, 2007).

9. G.P. Wallis & F. Jorge, 'Going under down under? Lineage ages argue for extensive survival of the Oligocene marine transgression on Zealandia', *Molecular Ecology 27*, 2018, pp. 4368–96.

10. M.A. Harper, B. Van de Vijver, U. Kaulfuss & D.E. Lee, 'Resolving the confusion between two fossil freshwater diatoms from Otago, New Zealand: *Encyonema jordanii* and *Encyonema jordaniforme* (Cymbellaceae, Bacillariophyta)', *Phytotaxa 394*, 2019, pp. 231–43.

11. D.C. Mildenhall, E.M. Kennedy, D.E. Lee et al., 'Palynology of the early Miocene Foulden Maar, Otago, New Zealand: Diversity following destruction', *RPP 204*, 2014, pp. 27–42.

12. J.K. Lindqvist & D.E. Lee, 'High-frequency paleoclimate signals from Foulden Maar, Waipiata Volcanic Field, southern New Zealand: An early Miocene varved lacustrine diatomite deposit', *Sedimentary Geology 222*, 2009, pp. 98–110.

13. Mildenhall, Kennedy, Lee et al., 'Palynology of the early Miocene Foulden Maar, Otago, New Zealand'; D.E. Lee, U. Kaulfuss, J.G. Conran et al., 'Biodiversity and palaeoecology of Foulden Maar: An early Miocene *Konservat-Lagerstätte* deposit in southern New Zealand', *Alcheringa 40*, 2016, pp. 525–41.

14. U. Kaulfuss, D.E. Lee, B.I.P Barratt et al., 'A diverse fossil terrestrial arthropod fauna from New Zealand: Evidence from the early Miocene Foulden Maar fossil *Lagerstätte*', *Lethaia 48*, 2014, pp. 299–308.

15. D.E. Lee, R.M. McDowall & J.K. Lindqvist, '*Galaxias* fossils from Miocene lake deposits, Otago, New Zealand: The earliest records of the Southern Hemisphere family Galaxiidae (Teleostei)', *JRSNZ 37*, 2007, pp. 109–30; U. Kaulfuss, D.E. Lee, J.H. Robinson et al., 'A review of *Galaxias* (Galaxiidae) fossils from the Southern Hemisphere', *Diversity 12*, 2020, p. 208; Kaulfuss, U., 'Geology and paleontology of Foulden Maar, Otago, New Zealand', PhD thesis, University of Otago, Dunedin, 2013.

16. Mildenhall, Kennedy, Lee et al., 'Palynology of the early Miocene Foulden Maar, Otago, New Zealand; Lee, Kaulfuss, Conran et al., 'Biodiversity and palaeoecology of Foulden Maar'.

17. L.J. Webb, 'A physiognomic classification of Australian rain forests', *Journal of Ecology 47*, 1959, pp. 551–70.

18. M. Pole, 'Plant macrofossils from the Foulden Hills Diatomite (Miocene), Central Otago, New Zealand', *JRSNZ 26*, 1996, pp. 1–39; D.E. Lee, J.M. Bannister, J.I. Raine & J.G. Conran, 'Euphorbiaceae: Acalyphoideae fossils from early Miocene New Zealand: *Mallotus–Macaranga* leaves, fruits, and inflorescence with *in situ Nyssapollenites endobalteus* pollen', *RPP 163*, 2010, pp. 127–38; J.G. Conran, J.M. Bannister, D.E. Lee et al., 'An update of monocot macrofossil data from New Zealand and Australia', *BJLS 178*, 2015, pp. 394–420; Lee, Kaulfuss, Conran et al., 'Biodiversity and palaeoecology of Foulden Maar'.

19. See Lee, Kaulfuss, Conran et al., 'Biodiversity and palaeoecology of Foulden Maar' and references therein.

20. Kaulfuss, Lee, Barratt et al., 'A diverse fossil terrestrial arthropod fauna from New Zealand'; U. Kaulfuss, D.E. Lee & A. Schmidt, 'New discoveries of Miocene arthropods from amber and diatomite deposits in New Zealand', in *7th International Conference on Fossil Insects, Arthropods and Amber*, D. Penney & A. Ross (eds), (Siri Scientific Press: 2016) p. 26.

21. A.C. Harris, J.M. Bannister & D.E. Lee, 'Fossil scale insects (Hemiptera, Coccoidea, Diaspididae) in life position on an angiosperm leaf from an early Miocene lake deposit, Otago, New Zealand', *JRSNZ 37*, 2007, pp. 1–13; Kaulfuss, Lee, Barratt et al., 'A diverse fossil terrestrial arthropod fauna from New Zealand'.

22. A.P. Kiecksee, U. Kaulfuss, D.E. Lee et al., 'Diversity of mites from New Zealand amber', in *Palaeobiology and Geobiology of Fossil Lagerstätten through Earth History*, J. Reitner, Q. Yang, Y. Wang & M. Reich (eds), (Göttingen: Universitätsdrucke Göttingen, 2013) pp. 86–87.

23. M.S. Engel & U. Kaulfuss, 'Diverse, primitive termites (Isoptera: Kalotermitidae, *incertae sedis*) from the early Miocene of New Zealand', *Austral Entomology 56*, 2017, pp. 94–103.

24. U. Kaulfuss, T. Wappler, E. Heiss & M.-C. Larivière, '*Aneurus* sp. from the early Miocene Foulden Maar, New Zealand: The first Southern Hemisphere record of fossil Aradidae (Insecta: Hemiptera: Heteroptera)', *JRSNZ 41*, 2011, pp. 279–85.

25. U. Kaulfuss, 'Geology and paleontology of Foulden Maar, Otago, New Zealand', PhD thesis, University of Otago, Dunedin, 2013.

26. J.G. Conran, W.G. Lee, D.E. Lee et al., 'Reproductive niche conservatism in the isolated New Zealand flora over 23 million years', *Biology Letters 10*, 2014, p. 20140647.

27. P.A. Selden & U. Kaulfuss, 'Fossil arachnids from the earliest Miocene Foulden Maar Fossil-Lagerstätte, New Zealand', *Alcheringa 43*, 2018, pp. 165–69.

28. T.H. Worthy, A.J.D. Tennyson, C. Jones et al., 'Miocene waterfowl and other birds from central Otago, New Zealand', *Journal of Systematic Palaeontology 5*, 2007, pp. 1–39; T.H. Worthy, S.J. Hand, J.P. Worthy et al., 'A large fruit pigeon (Columbidae) from the early Miocene of New Zealand', *The Auk 126*, 2009, pp. 649–56; T.H. Worthy, A.J.D. Tennyson & R.P. Scofield, 'An early Miocene diversity of parrots (Aves, Strigopidae, Nestorinae) from New Zealand', *Journal of Vertebrate Paleontology 31*, 2011, pp. 1102–16.

29. A. Driskell, L. Christidis, B.J. Gill et al., 'A new endemic family of New Zealand passerine birds: Adding heat to a biodiversity hotspot', *Australian Journal of Zoology 55*, 2007, pp. 73–78.

30. Kaulfuss, Lee, Barratt et al., 'A diverse fossil terrestrial arthropod fauna from New Zealand'.

31. U. Kaulfuss, D.E. Lee, J.-A. Wartho et al., 'Geology and palaeontology of the Hindon Maar Complex: A Miocene terrestrial fossil *Lagerstätte* in southern New Zealand', *PPP 500*, 2018, pp. 52–68.

32. U. Kaulfuss, D.E. Lee, J.H. Robinson et al., 'A review of *Galaxias* (Galaxiidae) fossils from the Southern Hemisphere', *Diversity 12*, 2020, p. 208.

33. J.D. Campbell, 'Casuarinaceae, Fagaceae, and other plant megafossils from Kaikorai Leaf Beds (Miocene) Kaikorai Valley, Dunedin, New Zealand', *NZJB 23*, 1985, pp. 311–20; T. Reichgelt, W.A. Jones, D.T. Jones et al., 'The flora of Double Hill (Dunedin Volcanic Complex, middle–late Miocene)', *JRSNZ 44*, 2014, pp. 105–35; Kaulfuss, Lee, Wartho et al., 'Geology and palaeontology of the Hindon Maar Complex'.

34. J.R. Nudds & P. Selden, *Extraordinary Fossil Ecosystems of North America* (London: Manson Publishing, 2008).

35. S. Schaal & W. Ziegler (eds), *Messel: An insight into the history of life and of the earth* (Oxford: Oxford University Press, 1992).

36. H. Lutz & U. Kaulfuss, 'A dynamic model for the meromictic lake Eckfeld Maar (Middle Eocene, Germany)', *Zeitschrift der Deutschen Gesellschaft für Geowissenschaften 157*, 2006, pp. 433–50; H. Lutz, U. Kaulfuss, T. Wappler et al., 'Eckfeld Maar: Window into an Eocene terrestrial habitat in Central Europe', *Acta Geologica Sinica 84*, 2010, pp. 984–1009.

37. K. Goth & P. Suhr, 'Die Forschungsbohrung Baruth 1998. Dokumentation der Entwicklungsgeschichte eines oberoligozänen Maarsees', *Geoprofil 12*, 2005, pp. 5–40; Goth & Suhr, *Baruths heiße Vergangenheit: Vulkane in der Lausitz* (Dresden: Sächsisches Landesamt für Umwelt und Geologie, 2007).

38. T.M. Kaiser, J. Ansorge, G. Arratia et al., 'The maar lake of Mahenge (Tanzania) – unique evidence of Eocene terrestrial environments in sub-Sahara Africa', *Zeitschrift der Deutschen Gesellschaft für Geowissenschaften 157*, 2006, pp. 411–31.

39. G. Liu & E.B. Leopold, 'Paleoecology of a Miocene flora from the Shanwang Formation, Shandong Province, Northern East China', *Palynology 16*, 1992, pp. 187–212; H. Yang & S. Yang, 'The Shanwang fossil biota in eastern China: A Miocene *Konservat-Lagerstätte* in lacustrine deposits', *Lethaia 27*, 1994, pp. 345–54.

40. B.W. Hayward, 'Best Practice Guide: Outstanding natural features. What are they and how should they be identified. How their significance might be assessed and documented?', *Geological Society of New Zealand Miscellaneous Publications 154*, 2019, pp. 1–26.

BIBLIOGRAPHY

Angiosperm Phylogeny Group, 'An update of the Angiosperm Phylogeny Group classification for the orders and families of flowering plants: APG III', *Botanical Journal of the Linnean Society 161*, 2009, pp. 105–21

Anonymous, 'Diatomaceous earth – discovery near Middlemarch', *ODT*, 6 July 1910, p. 4: https://paperspast. natlib.govt.nz/newspapers/ODT19100706.2.14

Arentsen, K., Németh, K. & Smid, E., *Abstracts volume of the Fourth International Maar Conference: A multidisciplinary congress on monogenetic volcanism. IAVCEI – CMV/ CVS – IAS 4IMC Conference Auckland, New Zealand 20–24 February 2012 (Geoscience Society of New Zealand Miscellaneous Publication 131A)* (Geoscience Society of New Zealand: Auckland, 2012)

Baker, R.G., Nest, J.V. & Woodworth, G., 'Dissimilarity coefficients for fossil pollen spectra from Iowa and western Illinois during the last 30,000 years', *Palynology 13*, 1989, pp. 63–77

Balme, B.E., 'Fossil *in situ* spores and pollen grains: An annotated catalogue', *Review of Palaeobotany and Palynology 87*, 1995, pp. 81–323

Bannister, J.M. & Conran, J.G., 'Comparative leaf morphology and cuticular anatomy of *Akania bidwillii* (Akaniaceae)', *Swainsona 33*, 2019, pp. 1–8

Bannister, J.M., Conran, J.G. & Lee, D.E., 'Life on the phylloplane: Eocene epiphyllous fungi from Pikopiko Fossil Forest, Southland, New Zealand', *New Zealand Journal of Botany 54*, 2016, pp. 412–32

Bannister, J.M., Lee, D.E. & Conran, J.G., 'Lauraceae from rainforest surrounding an early Miocene maar lake, Otago, southern New Zealand', *Review of Palaeobotany and Palynology 178*, 2012, pp. 13–34

Bannister, J.M., Lee, D.E. & Raine, J.I., 'Morphology and palaeoenvironmental context of *Fouldenia staminosa*, a fossil flower with associated pollen from the early Miocene of Otago, New Zealand', *New Zealand Journal of Botany 43*, 2005, pp. 515–25

Berry, P.E., Hahn, W.J., Sytsma, K.J., Hall, J.C. & Mast, A., 'Phylogenetic relationships and biogeography of *Fuchsia* (Onagraceae) based on noncoding nuclear and chloroplast DNA data', *American Journal of Botany 91*, 2004, pp. 601–14

Biffin, E., Hill, R.S. & Lowe, A.J., 'Did kauri (*Agathis*: Araucariaceae) really survive the Oligocene drowning of New Zealand?', *Systematic Biology 59*, 2010, pp. 594–602

Birch, J.L., Keeley, S.C. & Morden, C.W., 'Molecular phylogeny and dating of Asteliaceae (Asparagales): *Astelia s.l.* evolution provides insight into the Oligocene history of New Zealand', *Molecular Phylogenetics and Evolution 65*, 2012, pp. 102–15

Bolós, X., Oms, O., Rodríguez-Salgado, P., Martí, J., Gómez de Soler, B. & Campeny, G., 'Eruptive evolution and 3D geological modelling of Camp dels Ninots maar-diatreme (Catalonia) through continuous intra-crater drill coring', *Journal of Volcanology and Geothermal Research 419*, 2021, 107369.

Bowie, E., 'The geophysical characterization of Hindon Maar', PG Dip Sci thesis, University of Otago, Dunedin, 2015: http://theses.otagogeology.org.nz/items/show/593

Bradford, J.C., Hopkins, H.C.F. & Barnes, R.W., 'Cunoniaceae', in *The Families and Genera of Vascular Plants vol. 6. Flowering plants. Dicotyledons: Celastrales, Oxalidales, Rosales, Cornales, Ericales*, ed. K. Kubitzki (Berlin: Springer Verlag, 2004), pp. 91–111

Brea, M., Zucol, A.F., Bargo, M.S., Fernicola, J.C. & Vizcaíno, S.F., 'First Miocene record of Akaniaceae in Patagonia (Argentina): A fossil wood from the early Miocene Santa Cruz formation and its palaeobiogeographical implications', *Botanical Journal of the Linnean Society 183*, 2017, pp. 334–47

Breitwieser, I., Brownsey, P.J., Heenan, P.B., Nelson, W.A. & Wilton, A.D., *Flora of New Zealand Online*, 2010–21: www. nzflora.info

Buchanan, P.K., Beever, R.E., Glare, T.R., Johnston, P.R., McKenzie, E.H.C., Paulus, B.C., Pennycook, S.R., Ridley, G.S. & Smith, J.M.B., 'Kingdom Fungi: Introduction', in *New Zealand Inventory of Biodiversity vol. 3: Kingdoms Bacteria, Protozoa, Chromista, Plantae, Fungi*, D.P. Gordon (ed.) (Christchurch: Canterbury University Press, 2012), pp. 499–515

Büchel, G. & Wuttke, M., 'Reconstruction of the Eocene Messel Maar', in *IAVCEI – 5th International Maar Conference, Querétaro, México, Nov 17–22, 2014. Abstracts volume*, ed. Organising Committee (IAVCEI: Querétaro, Mexico, 2014), p. 93

Burnham, R.J., Ellis, B. & Johnson, K.R., 'Modern tropical forest taphonomy: Does high biodiversity affect paleoclimatic interpretations?', *Palaios 20*, 2005, pp. 439–51

Campbell, H., *The Zealandia Drowning Debate: Did New Zealand sink beneath the waves?* (Wellington: Bridget Williams Books, 2013)

Campbell, H. & Hutching, G., *In Search of Ancient New Zealand* (Auckland: Penguin, 2007)

Campbell, J.D., 'Casuarinaceae, Fagaceae, and other plant megafossils from Kaikorai Leaf Beds (Miocene) Kaikorai Valley, Dunedin, New Zealand', *New Zealand Journal of Botany 23*, 1985, pp. 311–20

Carpenter, R.J., Bannister, J.M., Lee, D.E. & Jordan, G.J., 'Proteaceae leaf fossils from the Oligo-Miocene of New Zealand: New species and evidence of biome and trait conservatism', *Australian Systematic Botany 25*, 2012, pp. 375–89

Carpenter, R.J., Jordan, G.J., MacPhail, M.K. & Hill, R.S., 'Near-tropical early Eocene terrestrial temperatures at the Australo-Antarctic margin, western Tasmania', *Geology 40*, 2012, pp. 267–70

Carpenter, R.J., Wilf, P., Conran, J.G. & Cúneo, N.R., 'A Paleogene trans-Antarctic distribution for *Ripogonum* (Ripogonaceae: Liliales)?', *Palaeontologia Electronica 17*, 2014, pp. 39A, 1–9

Chase, M.W., Christenhusz, M.J.M. & Mirenda, T., 'Orchid evolution', in *The Book of Orchids: A life-size guide to six hundred species from around the world*, M.W. Chase, M.J.M. Christenhusz & T. Mirenda (eds) (Chicago: University of Chicago Press, 2017), pp. 12–13

Christenhusz, M.J.M. & Byng, J.W., 'The number of known plants species in the world and its annual increase', 2016, pp. 201–17

Christenhusz, M.J.M., Fay, M.F. & Chase, M.W., *Plants of the World: An illustrated encyclopedia of vascular plants* (Richmond, UK: RBG Kew and Chicago: University of Chicago Press, 2017)

Christie, A.B. & Barker, R.G., *Mineral Resource Assessment of the Northland Region, New Zealand*, GNS Science Report 2007/06 (Lower Hutt: GNS Science, 2007)

Cieraad, E. & Lee, D.E., 'The New Zealand fossil record of ferns for the past 85 million years', *New Zealand Journal of Botany 44*, 2006, pp. 143–70

Conran, J.G., '*Paracordyline kerguelensis*, an Oligocene monocotyledon macrofossil from the Kerguélen Islands', *Alcheringa 21*, 1997, pp. 129–40

Conran, J.G., Bannister, J.M. & Lee, D.E., 'Earliest orchid macrofossils: Early Miocene *Dendrobium* and *Earina* (Orchidaceae: Epidendroideae) from New Zealand', *American Journal of Botany 96*, 2009, pp. 466–74

Conran, J.G., Bannister, J.M. & Lee, D.E., 'Fruits and leaves with cuticle of *Laurelia otagoensis sp. nov.* (Atherospermataceae) from the early Miocene of Otago (New Zealand)', *Alcheringa 37*, 2013a, pp. 496–509

Conran, J.G., Bannister, J.M. & Lee, D.E., 'Fossil fruits and seeds from the early Miocene Foulden Maar, New Zealand', in *Geosciences 2013, Geoscience Society of New Zealand annual conference, 24–27 Nov 2013*, GSNZ Miscellaneous Publications 136A, C.M. Reid & A. Wandres (eds) (Christchurch: GSNZ, 2013b) p. 20

Conran, J.G., Bannister, J.M., Lee, D.E., Carpenter, R.J., Kennedy, E.M., Reichgelt, T. & Fordyce, R.E., 'An update of monocot macrofossil data from New Zealand and Australia', *Botanical Journal of the Linnean Society 178*, 2015, pp. 394–420

Conran, J.G., Bannister, J.M., Mildenhall, D.C., Lee, D.E., Chacón, J. & Renner, S.S., 'Leaf fossils of *Luzuriaga* and a monocot flower with in situ pollen of *Liliacidites contortus* Mildenh. & Bannister sp. nov. (Alstroemeriaceae) from the early Miocene', *American Journal of Botany 101*, 2014, pp. 141–55

Conran, J.G., Bannister, J.M., Reichgelt, T. & Lee, D.E., 'Epiphyllous fungi and leaf physiognomy indicate an ever-wet humid mesothermal (subtropical) climate in the late Eocene of southern New Zealand', *Palaeogeography, Palaeoclimatology, Palaeoecology 452*, 2016, pp. 1–10

Conran, J.G., Carpenter, R.J. & Jordan, G.J., 'Early Eocene *Ripogonum* (Liliales: Ripogonaceae) leaf macrofossils from southern Australia', *Australian Systematic Botany 22*, 2009, pp. 219–28

Conran, J.G. & Christophel, D.C., '*Paracordyline aureonemoralis* (Lomandraceae): An Eocene monocotyledon from South Australia', *Alcheringa 22*, 1998, pp. 351–59

Conran, J.G., Jackson, J.A., Lee, D.E. & Kennedy, E.M., '*Gleichenia*-like *Korallipteris alineae* sp. nov. macrofossils (Polypodiophyta) from the Miocene Landslip Hill silcrete, New Zealand', *New Zealand Journal of Botany 55*, 2017, pp. 258–75

Conran, J.G., Kaulfuss, U., Bannister, J.M., Mildenhall, D.C. & Lee, D.E., '*Davallia* (Polypodiales: Davalliaceae) macrofossils from early Miocene Otago (New Zealand) with in situ spores', *Review of Palaeobotany and Palynology 162*, 2010, pp. 84–94

Conran, J.G., Kaulfuss, U., Bannister, J.M., Mildenhall, D.C. & Lee, D.E., 'An *Akania* (Akaniaceae) inflorescence with associated pollen from the early Miocene of New Zealand', *American Journal of Botany 106*, 2019, pp. 1–11

Conran, J.G., Kennedy, E.M. & Bannister, J.M., 'Early Eocene Ripogonaceae leaf macrofossils from New Zealand', *Australian Systematic Botany 31*, 2018, pp. 8–15

Conran, J.G., Lee, D.E. & Bannister, J.M., 'Kiwi Curare: Miocene Menispermaceae in New Zealand', in *Southern Lands and Southern Oceans: Life on the edge? VII Southern Connection Congress, Dunedin NZ, 21–25 Jan 2013*, SC Editorial Committee (ed.) (Dunedin: SC Editorial Committee, 2013) p. 24

Conran, J.G., Lee, W.G., Lee, D.E., Bannister, J.M. & Kaulfuss, U., 'Reproductive niche conservatism in the isolated New Zealand flora over 23 million years', *Biology Letters 10*, 2014, p. 20140647

Conran, J.G., Mildenhall, D.C., Raine, J.I., Kennedy, E.M. & Lee, D.E., 'The monocot fossil pollen record of New Zealand and its implications for palaeoclimates and environments', *Botanical Journal of the Linnean Society 178*, 2015, pp. 421–40

Coode, M.J.E., 'Elaeocarpaceae', in *The Families and Genera of Vascular Plants vol. 6. Flowering plants. Dicotyledons: Celastrales, Oxalidales, Rosales, Cornales, Ericales*, K. Kubitzki (ed.) (Berlin: Springer Verlag, 2004), pp. 135–44

Coombs, D.S., Adams, C.J., Roser, B.P. & Reay, A., 'Geochronology and geochemistry of the Dunedin Volcanic Group, eastern Otago, New Zealand', *New Zealand Journal of Geology and Geophysics 51*, 2008, pp. 195–218

Cooper, W. & Cooper, W.T., *Fruits of the Australian Tropical Rainforest* (Melbourne: Nokomis Editions, 2004)

Couper, R.A., 'Plant microfossils from the Middlemarch diatomite (Appendix to Coombs et al.: Age relations of the Dunedin Volcanic Complex and some paleogeographic implications, part 2)', *New Zealand Journal of Geology and Geophysics 3*, 1960, pp. 578–79

D'Andrea, W.J., Donatich, S., Fox, B.R.S. & Lee, D.E., 'Orbital pacing of New Zealand hydroclimate during the Early Miocene: Biomarker and compound-specific isotope records from the Foulden Maar diatomite', in *Southern Lands and Southern Oceans: Life on the edge? VII Southern Connection Congress, Dunedin NZ, 21–25 Jan 2013*, SC Editorial Committee (ed.) (Dunedin: SC Editorial Committee, 2013)

D'Andrea, W.J., Fox, B.R.S. & Lee, D.E., 'Interannual and orbital-scale climate variability in the early Miocene: Compound-specific D/H records from the Foulden Maar Diatomite, New Zealand', in *American Geophysical Union, Fall Meeting, San Francisco, 15–19 Dec 2014, abstracts*, Organising Committee (ed.) (AGU: 2014) pp. PP43B–2095

Darwin, C., *The Origin of Species by Means of Natural Selection*, 6th (reprint) edn (London: J.M. Dent & Sons Ltd, 1872)

Dawson, J. & Lucas, R., *New Zealand's Native Trees* (Nelson: Craig Potton Publishing, 2011)

de Jong, H., Oosterbroek, P., Gelhaus, J., Reusch, H. & Young, C., 'Global diversity of craneflies (Insecta, Diptera: Tipulidea or Tipulidae sensu lato) in freshwater', in *Freshwater Animal Diversity Assessment: Developments in hydrobiology, vol. 198*, E.V. Balian, C. Lévêque, H. Segers & K. Martens (eds) (Dordrecht: Springer, 2007)

Dettmann, M.E. & Clifford, H.T., 'The fossil record of *Elaeocarpus* L. fruits', *Memoirs of the Queensland Museum 46*, 2000, pp. 461–97

Don, W., *Ants of New Zealand* (Dunedin: Otago University Press, 2007)

Donatich, S., D'Andrea, W.J., Fox, B.R.S. & Lee, D.E., 'Orbital pacing of New Zealand hydroclimate during the Early Miocene: Biomarker and compound-specific isotope records from the Foulden Maar diatomite', in *American Geophysical Union, Fall Meeting, San Francisco, 3–7 Dec 2012, abstracts*, Organising Committee (ed.) (AGU: 2012) pp. PP11C–2040

Driskell, A., Christidis, L., Gill, B.J., Boles, W.E., Barker, F.K. & Longmore, N.W., 'A new endemic family of New Zealand passerine birds: Adding heat to a biodiversity hotspot', *Australian Journal of Zoology 55*, 2007, pp. 73–78

Engel, M.S. & Kaulfuss, U., 'Diverse, primitive termites (Isoptera: Kalotermitidae, *incertae sedis*) from the early Miocene of New Zealand', *Austral Entomology 56*, 2017, pp. 94–103

Evans, K., 'Ancient "pickled" leaves give a glimpse of global greening', *Eos 101*, 2020: https://eos.org/articles/ancient-pickled-leaves-give-a-glimpse-of-global-greening

Evans, K., 'In New Zealand, locals rally to save fossils from destruction', *Undark*, 29 Mar 2021a: https://undark.org/2021/03/29/new-zealand-save-fossils-from-destruction/

Evans, K., 'New Zealanders are rallying to save local fossils from destruction', *Atlas Obscura*, 6 Apr 2021b: www.atlasobscura.com/articles/new-zealand-fossils

Faegri, K. & van der Pijl, L., *The Principles of Pollination Ecology*, 3rd (revised) edn (Oxford: Pergamon, 1979)

Fairburn, C., 'Mining black pearl', *Quarrying and Mining Magazine*, 13 Aug 2018: https://quarryingandminingmag.co.nz/mining-black-pearl/

Forbes, A.A., Bagley, R.K., Beer, M.A., Hippee, A.C. & Widmayer, H.A., 'Quantifying the unquantifiable: Why Hymenoptera, not Coleoptera, is the most speciose animal order', *BMC Ecology 18*, 2018, p. 21

Forster, J.R. & Forster, G., *Characteres generum plantarum quas in itinere ad insulas Maris Australis, collegerunt, descripserunt, delinearunt, annis MDCCLXXII–MDCCLXXV*

/ *Joannes Reinoldus Forster et Georgius Forster*, 2nd edn (Londini: Prostant apud B. White, T. Cadell & P. Elmsly, 1776)

Forsyth, P.J. (ed.), *Geology of the Waitaki Area 1:250 000 geological map 19* (Lower Hutt: Institute of Geological & Nuclear Sciences, 2001)

Foster, G.L., Royer, D.L. & Lunt, D.J., 'Future climate forcing potentially without precedent in the last 420 million years', *Nature Communications 8*, 2017, p. 14845

Fox, B.R.S., 'Climate change at the Oligocene/Miocene boundary', PhD thesis, University of Otago, Dunedin, 2014

Fox, B.R.S., D'Andrea, W.J., Wilson, G.S. & Lee, D.E., 'Tropical influence on Miocene New Zealand', in *Geosciences 2014. Annual conference of the Geoscience Society of New Zealand, 24–27 Nov 2014, New Plymouth: Abstracts. GSNZ Miscellaneous Publication 139A*, Organising Committee (ed.) (New Plymouth, GSNZ, 2014), p. 38

Fox, B.R.S., D'Andrea, W.J., Wilson, G.S., Lee, D.E. & Wartho, J.-A., 'Interaction of polar and tropical influences in the mid-latitudes of the Southern Hemisphere during the Mi-1 deglaciation', *Global and Planetary Change 155*, 2017, pp. 109–20

Fox, B.R.S., Haworth, M., Wilson, G., Lee, D.E., Wartho, J.A., Bannister, J.M., Jones, D., Kaulfuss, U. & Lindqvist, J.K., 'Greenhouse gases and deglaciation at the Oligocene/Miocene boundary: Palaeoclimate evidence from a New Zealand maar lake', in *Abstracts volume of the Fourth International Maar Conference: A multidisciplinary congress on monogenetic volcanism. IAVCEI – CMV/CVS – IAS 4IMC Conference Auckland, New Zealand, 20–24 Feb 2012 (Geoscience Society of New Zealand Miscellaneous Publication 131A)*, K. Arentsen, K. Németh & E. Smid (eds) (Auckland: Geoscience Society of New Zealand, 2012) p. 29

Fox, B.R.S., Wartho, J., Wilson, G.S., Lee, D.E., Nelson, F.E. & Kaulfuss, U., 'Long-term evolution of an Oligocene/Miocene maar lake from Otago, New Zealand', *Geochemistry, Geophysics, Geosystems 16*, 2015, pp. 59–76

Fox, B.R.S., Wilson, G.S. & Lee, D.E., 'A unique annually laminated maar lake sediment record shows orbital control of Southern Hemisphere mid-latitude climate across the Oligocene–Miocene boundary', *Geological Society of America Bulletin 128*, 2016, pp. 609–26

Fox, B.R.S., Wilson, G.S., Lee, D.E., Haworth, M., Wartho, J., Kaulfuss, U., Bannister, J.M., Gorman, A.R., Jones, D.A. & Lindqvist, J.K., 'Atmospheric carbon dioxide as a driver for deglaciation during the Mi-1 event: New evidence from terrestrial Southern Hemisphere proxies (abstract #PP13A-1806)', in *American Geophysical Union, Fall Meeting 5–9 Dec 2011*, American Geophysical Union

(ed.) (American Geophysical Union: 2011): www.agu.org/meetings/abstract_db.shtml

Fox, B.R.S., Kaulfuss, U., Jones, D., Wilson, G.S., Lee, D.E. & Gorman, A.R., 'A Miocene terrestrial sediment core from Foulden Maar, Otago', in *Joint Geological and Geophysical Societies Conference, Oamaru, 23–27 Nov 2009: Programme and abstracts. Geological Society of New Zealand, Miscellaneous Publication 128A*, D.J.A. Barrell & A. Tulloch (eds) (Ōamaru: GSNZ, 2009) p. 73

Fox, B.R.S., Wartho, J., Wilson, G.S., Lee, D.E., Nelson, F.E. & Kaulfuss, U., 'Long-term evolution of an Oligocene/Miocene maar lake from Otago, New Zealand', *Geochemistry, Geophysics, Geosystems 16*, 2015, pp. 59–76

Fox, B.R.S., Wilson, G.S. & Lee, D.E., 'A unique annually laminated maar lake sediment record shows orbital control of Southern Hemisphere mid-latitude climate across the Oligocene–Miocene boundary', *Geological Society of America Bulletin 128*, 2016, pp. 609–26

Fox, B.R.S., Wilson, G.S. & Lee, D.E., 'Mi-1 deglaciation characterised by abrupt short-term cooling events', in *Southern Lands and Southern Oceans: Life on the edge? VII Southern Connection Congress, Dunedin NZ, 21–25 Jan 2013*, SC Editorial Committee (ed.) (Dunedin: SC Editorial Committee, 2013)

Fox, B.R.S., Wilson, G.S., Lee, D.E., Haworth, M., Wartho, J., Bannister, J.M. & Kaulfuss, U., 'Causes and consequences of climate change during the Mi-1 Event: Paleoclimate data from the Foulden Maar core', in *Geosciences 2011. Geoscience Society of New Zealand annual conference, 27 Nov–1 Dec 2011*, GSNZ Miscellaneous Publications 130A, N.J. Litchfield & K. Clark (eds) (Nelson: GSNZ, 2011) p. 39

Friis, E.M., Crane, P.R. & Pedersen, K.R., *Early Flowers and Angiosperm Evolution* (Cambridge: Cambridge University Press, 2011)

Gandolfo, M.A., Dibbern, M.C. & Romero, E.J., '*Akania patagonica n. sp.* and additional material on *Akania americana* Romero & Hickey (Akaniaceae), from Paleocene sediments of Patagonia', *Bulletin of the Torrey Botanical Club 115*, 1988, pp. 83–88

Gee, C.T., 'The genesis of mass carpological deposits (Bedload Carpodeposits) in the Tertiary of the lower Rhine Basin, Germany', *Palaios 20*, 2005, pp. 463–78

Gerrard, J., 'Digging for the truth on fossils, profit, and the Foulden Maar mine (Originally published at PMCSA.ac.nz)', *The Spinoff*, 20 May 2019: https://thespinoff.co.nz/author/juliet-gerrard/

Gibbs, George W., *Ghosts of Gondwana: The history of life in New Zealand*, rev. edn (Nelson: Craig Potton Publishing, 2016)

Godley, E.J. & Berry, P.E., 'The biology and systematics of *Fuchsia* in the South Pacific', *Annals of the Missouri Botanical Garden 82*, 1995, pp. 473–516

Gómez de Soler, B., Campeny Vall-Llosera, G., Van der Made, J., Oms, O., Agustí, J., Sala, R., Blain, H.-A., Burjachs, F., Claude, J., García Catalán, S., Riba, D. & Rosillo, R., 'A new key locality for the Pliocene vertebrate record of Europe: The Camp dels Ninots maar (NE Spain)', *Geologica Acta 10*, 2012, pp. 1–17

Gordon, F.R., *The Middlemarch Diatomite Deposit: The completion of the exploratory programme and development* (unpublished mining report) (Dunedin: School of Mines and Metallurgy, 1954)

Gordon, F.R., 'Paper 136: The occurrence of diatomite near Middlemarch, Otago', in *Proceedings of the Fourth Triennial Mineral Conference, School of Mines and Metallurgy, University of Otago, Dunedin, New Zealand, 1–3 September 1959* (Dunedin: University of Otago, 1959a), p. 9

Gordon, F.R., 'Paper 137: The properties and uses of Middlemarch diatomite', in *Proceedings of the Fourth Triennial Mineral Conference, School of Mines and Metallurgy, University of Otago, Dunedin, New Zealand, 1–3 September 1959* (Dunedin: University of Otago, 1959b), p. 13

Goth, K. & Suhr, P., 'Die Forschungsbohrung Baruth 1998. Dokumentation der Entwicklungsgeschichte eines oberoligozänen Maarsees', *Geoprofil 12*, 2005, pp. 5–40

Goth, K. & Suhr, P., *Baruths heiße Vergangenheit: Vulkane in der Lausitz* (Dresden: Sächsisches Landesamt für Umwelt und Geologie, 2007)

Graham, I.J. (chief ed.), *A Continent on the Move: New Zealand geoscience revealed*, 2nd edn., Geoscience Society of New Zealand Miscellaneous Publication 141 (Wellington: Geoscience Society of New Zealand, 2015)

Grange, L.I., 'Diatomite: Principal New Zealand sources and uses', *New Zealand Journal of Science and Technology 12*, 1930, pp. 94–99

Gravis, I., Németh, K., Twemlow, C. & Németh, B., 'The case for community-led geoheritage and geoconservation ventures in Māngere, South Auckland, and Central Otago, New Zealand', *Geoheritage 12*, 2020, p. 19

Greenwood, D.R., 'Leaf margin analysis: Taphonomic constraints', *Palaios 20*, 2005, pp. 498–505

Grein, M., Konrad, W., Wilde, V., Utescher, T. & Roth-Nebelsick, A., 'Reconstruction of atmospheric CO_2 during the early middle Eocene by application of a gas exchange model to fossil plants from the Messel Formation, Germany', *Palaeogeography, Palaeoclimatology, Palaeoecology 309*, 2011, pp. 383–91

Grímsson, F., Zetter, R., Labandeira, C.C., Engel, M.S., Wappler, T., 'Taxonomic description of *in situ* bee pollen from the middle Eocene of Germany', *Grana 56*, 2017, pp. 37–70

Hancock, F., 'Fossil-dirt nutrition claims under doubt', *Newsroom*, 17 May 2019: www.newsroom.co.nz/fossil-dirt-nutrition-claims-under-doubt

Hansen, J., Sato, M., Russell, G. & Kharecha, P., 'Climate sensitivity, sea level and atmospheric carbon dioxide', *Philosophical Transactions of the Royal Society A: Mathematical, Physical and Engineering Sciences 371*, 2013, pp. 20120294

Hansford, D., 'Buried treasure', *New Zealand Geographic*, July 2019: www.nzgeo.com/stories/buried-treasure/

Harper, M.A., van de Vijver, B., Kaulfuss, U. & Lee, D.E., 'Resolving the confusion between two fossil freshwater diatoms from Otago, New Zealand: *Encyonema jordanii* and *Encyonema jordaniforme* (Cymbellaceae, Bacillariophyta)', *Phytotaxa 394*, 2019, pp. 231–43

Harris, A.C., 'An Eocene larval insect fossil (Diptera Bibionidae) from North Otago, New Zealand', *Journal of the Royal Society of New Zealand 13*, 1983, pp. 93–105

Harris, A.C., Bannister, J.M. & Lee, D.E., 'Fossil scale insects (Hemiptera, Coccoidea, Diaspididae) in life position on an angiosperm leaf from an early Miocene lake deposit, Otago, New Zealand', *Journal of the Royal Society of New Zealand 37*, 2007, pp. 1–13

Hartley, S., 'Otago fertiliser to bolster oil palm plantations', *ODT*, 13 Nov 2015: www.odt.co.nz/business/otago-fertiliser-bolster-oil-palm-plantations

Hartley, S., 'Potential to create 100 jobs', *ODT*, 28 May 2018: www.odt.co.nz/news/dunedin/potential-create-100-jobs

Hartley, S., 'Funding secured by diatomite mine firm', *ODT*, 18 Jun 2018: www.odt.co.nz/business/funding-secured-diatomite-mine-firm

Hartley, S., 'Leaked report sheds light on mine project', *ODT*, 20 Apr 2019: www.odt.co.nz/business/leaked-report-sheds-light-mine-project

Hayward, B.W., *Best Practice Guide: Outstanding natural features. What are they and how should they be identified? How their significance might be assessed and documented?*, Geological Society of New Zealand Miscellaneous Publications 154, 2019a, pp. 1–26

Hayward, B.W., *Volcanoes of Auckland: A field guide* (Auckland: Auckland University Press, 2019b)

Hocken, A.G., 'The early life of James Hector, 1834 to 1865: The first Otago Provincial Geologist', PhD thesis, University of Otago, Dunedin, 2008

Holden, A.M., 'Studies in New Zealand Oligocene and Miocene plant macrofossils', PhD thesis, Victoria University, Wellington, 1983.

Homes, A.M., Cieraad, E., Lee, D.E., Raine, J.I. & Conran, J.G., 'A diverse fern flora including macrofossils with *in situ* spores from the Late Eocene of southern New Zealand', *Review of Palaeobotany and Palynology 220*, 2015, pp. 16–28

Hunter, J.A., 'A further note on *Tecomanthe speciosa* W.R.B. Oliver (Bignoniaceae)', *Records of the Auckland Institute and Museum 6*, 1967, pp. 169–70

Hutton, F.W. & Ulrich, G.H.F., *Report on the Geology and Gold Fields of Otago* (Dunedin: Provincial Council Otago, 1875)

Jellyman, D.J., 'Management and fisheries of Australasian eels (*Anguilla australis, Anguilla dieffenbachii, Anguilla reinhardtii*)', in *Biology and Ecology of Anguillid Eels*, T. Arai (ed.) (London: CRC Press, 2016), pp. 274–90

Jenkins Shaw, J., Solodovnikov, A., Bai, M. & Kaulfuss, U., 'An amblyopinine rove beetle (Coleoptera, Staphylinidae, Staphylininae, Amblyopinini) from the earliest Miocene Foulden Maar fossil-Lagerstätte, New Zealand', *Journal of Paleontology 94*, 2020, pp. 1082–88

Jérémie, J., 'Étude des Monimiaceae: Révision du genre *Hedycarya*', *Adansonia, séries 2 18*, 1978, pp. 25–53

Jones, D.A., 'The geophysical characterisation of the Foulden Maar', MSc thesis, University of Otago, Dunedin, 2012: http://hdl.handle.net/10523/2334

Jones, D.A., Wilson, G.S., Gorman, A.R., Fox, B.R.S., Lee, D.E. & Kaulfuss, U., 'A drill-hole calibrated geophysical characterisation of the 23 Ma Foulden Maar stratigraphic sequence, Otago, New Zealand', *New Zealand Journal of Geology and Geophysics 60*, 2017, pp. 465–77

Jordan, G.J., Carpenter, R.J., Bannister, J.M., Lee, D.E., Mildenhall, D.C. & Hill, R.S., 'High conifer diversity in Oligo–Miocene New Zealand', *Australian Systematic Botany 24*, 2011, pp. 121–36

Jouzel, J., Masson-Delmotte, V., Cattani, O., Dreyfus, G., Falourd, S., Hoffmann, G., Minster, B., Nouet, J., Barnola, J.M., Chappellaz, J., Fischer, H., Gallet, J.C., Johnsen, S., Leuenberger, M., Loulergue, L., Luethi, D., Oerter, H., Parrenin, F., Raisbeck, G., Raynaud, D., Schilt, A., Schwander, J., Selmo, E., Souchez, R., Spahni, R., Stauffer, B., Steffensen, J.P., Stenni, B., Stocker, T.F., Tison, J.L., Werner, M. & Wolff, E.W., 'Orbital and millennial Antarctic climate variability over the past 800,000 years', *Science 317*, 2007, pp. 793–96

Kaiser, T.M., Ansorge, J., Arratia, G., Bullwinkel, V., Gunnell, G.F., Herendeen, P.S., Jacobs, B., Mingram, J., Msuya, C., Musolff, A., Naumann, R., Schulz, E. & Wilde, V., 'The maar lake of Mahenge (Tanzania) – unique evidence of Eocene terrestrial environments in sub-Sahara Africa', *Zeitschrift der Deutschen Gesellschaft für Geowissenschaften 157*, 2006, pp. 411–31

Kamp, P.J.J., Vincent, K.A. & Tayler, M.J.S., *Cenozoic Sedimentary and Volcanic Rocks of New Zealand: A reference volume of lithology, age and paleoenvironments with maps (PMAPs) and database* (Hamilton: University of Waikato, 2015)

Kaulfuss, U., 'Geology and paleontology of Foulden Maar, Otago, New Zealand', PhD thesis, University of Otago, Dunedin, 2013

Kaulfuss, U., 'Crater stratigraphy and the post-eruptive evolution of Foulden Maar, southern New Zealand', *New Zealand Journal of Geology and Geophysics 60*, 2017, pp. 410–32

Kaulfuss, U., Brown, S.D.J., Henderson, I.M., Szwedo, J. & Lee, D.E., 'First insects from the Manuherikia Group, early Miocene, New Zealand', *Journal of the Royal Society of New Zealand 49*, 2018, pp. 494–507

Kaulfuss, U., Conran, J.G., Bannister, J.M., Mildenhall, D.C. & Lee, D.E., 'A new Miocene fern (*Palaeosorum*: Polypodiaceae) from New Zealand bearing in situ spores of *Polypodiisporites*', *New Zealand Journal of Botany 57*, 2019, pp. 2–17

Kaulfuss, U. & Dlussky, G.M., 'Early Miocene Formicidae (Amblyoponinae, Ectatomminae, ?Dolichoderinae, Formicinae, and Ponerinae) from the Foulden Maar Fossil Lagerstätte, New Zealand, and their biogeographic relevance', *Journal of Paleontology 89*, 2015, pp. 1043–55

Kaulfuss, U., Harris, A.C. & Lee, D.E., 'A new fossil termite (Isoptera, Stolotermitidae, *Stolotermes*) from the early Miocene of Otago, New Zealand', *Acta Geologica Sinica 84*, 2010, pp. 705–09

Kaulfuss, U., Harris, A.C., Conran, J.G. & Lee, D.E., 'An early Miocene ant (subfam. Amblyoponinae) from Foulden Maar: The first fossil Hymenoptera from New Zealand', *Alcheringa 38*, 2014, pp. 568–74

Kaulfuss, U. & Lee, D.E., 'Foulden Maar: A new locality for fossil insects and spiders from the early Miocene of New Zealand', in *The 5th International Conference on Fossil Insects, Arthropods and Amber, Beijing, China, 20–25 Aug 2010, abstract volume*, Organising Committee (ed.) (Fossils X3: 2010) p. 144

Kaulfuss, U., Lee, D.E., Barratt, B.I.P., Leschen, R.A.B., Larivière, M.-C., Dlussky, G.M., Henderson, I.M. & Harris, A.C., 'A diverse fossil terrestrial arthropod fauna from New Zealand: evidence from the early Miocene Foulden Maar fossil *Lagerstätte*', *Lethaia 48*, 2014, pp. 299–308

Kaulfuss, U., Lee, D.E., Robinson, J.H., Wallis, G.P. & Schwarzhans, W.W., 'A review of *Galaxias* (Galaxiidae) fossils from the Southern Hemisphere', *Diversity 12*, 2020, p. 208

Kaulfuss, U., Lee, D.E. & Schmidt, A., 'New discoveries of Miocene arthropods from amber and diatomite deposits in New Zealand', in *7th International conference on fossil insects, arthropods and amber, Edinburgh, 26 Apr–1 May 2016. Abstracts*, D. Penney & A. Ross (eds) (Siri Scientific Press: 2016) p. 26

Kaulfuss, U., Lee, D.E. & Wappler, T., 'The Foulden Maar diatomite – insight into an Early Miocene terrestrial ecosystem of New Zealand', in *Paläontologie. Schlüssel zur Evolution. 79. Jahrestagung der Paläontologischen Gesellschaft, Bonn, 5–7 Oct 2009. Terra Nostra 60–79*, T. Martin & S.I. Kaiser (eds) (Potsdam: GeoUnion Alfred-Wegener-Stiftung, 2009)

Kaulfuss, U., Lee, D.E., Wartho, J.-A., Bowie, E., Lindqvist, J.K., Conran, J.G., Bannister, J.M., Mildenhall, D.C., Kennedy, E.M. & Gorman, A.R., 'Geology and palaeontology of the Hindon Maar Complex: A Miocene terrestrial fossil *Lagerstätte* in southern New Zealand', *Palaeogeography, Palaeoclimatology, Palaeoecology 500*, 2018, pp. 52–68

Kaulfuss, U., Lindqvist, J.K., Jones, D., Fox, B.R.S., Wilson, G. & Lee, D.E., 'Post-eruptive maar crater sedimentation inferred from outcrop, drill cores and geophysics – Foulden Maar, Early Miocene, Waipiata Volcanic Field, New Zealand', in *Abstracts Volume of the Fourth International Maar Conference: A multidisciplinary congress on monogenetic volcanism. IAVCEI – CMV/CVS – IAS 4IMC Conference Auckland, New Zealand 20–24 February 2012 (Geoscience Society of New Zealand Miscellaneous Publication 131A)*, K. Arentsen, K. Németh & E. Smid (eds) (Auckland: Geoscience Society of New Zealand, 2012) p. 43

Kaulfuss, U. & Moulds, M., 'A new genus and species of tettigarctid cicada from the early Miocene of New Zealand: *Paratettigarcta zealandica* (Hemiptera, Auchenorrhyncha, Tettigarctidae)', *ZooKeys 484*, 2015, pp. 83–94

Kaulfuss, U., Wappler, T., Heiss, E. & Larivière, M.-C., '*Aneurus* sp. from the early Miocene Foulden Maar, New Zealand: the first Southern Hemisphere record of fossil Aradidae (Insecta: Hemiptera: Heteroptera)', *Journal of the Royal Society of New Zealand 41*, 2011, pp. 279–85

Kelly, M., Edwards, A.R., Wilkinson, M.R., Alvarez, B., Cook, S.d.C., Bergquist, P.R., Buckeridge, J.S., Campbell, H., Reiswig, H.M. & Valentine, C., 'Phylum Porifera sponges', in *New Zealand Inventory of Biodiversity vol. 1: Kingdom Animalia: Radiata, Lophotrochozoa and Deuterostomia*, ed. D.P. Gordon (Christchurch: Canterbury University Press, 2009), pp. 23–46

Kelly, D., Ladley, J.J. & Robertson, A.W., 'Is dispersal easier than pollination? Two tests in New Zealand Loranthaceae', *New Zealand Journal of Botany 42*, 2004, pp. 89–103

Kelly, D., Ladley, J.J., Robertson, A.W., Anderson, S.H., Wotton, D.M. & Wiser, S.K., 'Mutualisms with the wreckage of an avifauna: The status of bird pollination and fruit dispersal in New Zealand', *New Zealand Journal of Ecology 34*, 2010, pp. 66–85

Kennedy, E.M., Lovis, J.D. & Daniel, I.L., 'Discovery of a Cretaceous angiosperm reproductive structure from New Zealand', *New Zealand Journal of Geology and Geophysics 46*, 2003, pp. 519–22

Kerr, I.A., Conran, J.G., Lee, D.E. & Waycott, M., 'Phylogeny, fossil history and biogeography of Ripogonaceae', in *Abstracts, Geosciences 2016, Wanaka, Geoscience Society of New Zealand annual conference, 28 Nov–1 Dec 2016. GSNZ Miscellaneous Publication 145A*, C. Riesselmann & A. Roben (eds) (Wanaka: GSNZ, 2016), p. 103

Kessler, P.J.A., 'Menispermaceae', in *The Families and Genera of Vascular Plants vol. 2. Flowering plants. Dicotyledons: Magnoliid, hamamelid and caryophyllid families*, K. Kubitzki, J.G. Rohwer & V. Bittrich (eds) (Berlin: Springer Verlag, 1993), pp. 402–18

Kiecksee, A.P., Kaulfuss, U., Lee, D.E., Sadowski, E.-M., Schmidt, A.R. & Maraun, M., 'Diversity of mites from New Zealand amber', in *Palaeobiology and Geobiology of Fossil Lagerstätten through Earth History: A joint conference of the 'Paläontologische Gesellschaft' and the 'Palaeontological Society of China', Göttingen, Germany, Sep 23–27, 2013. Abstract Volume*, J. Reitner, Q. Yang, Y. Wang & M. Reich (eds) (Göttingen: Universitätsdrucke Göttingen, 2013) pp. 86–87

King, P., Milicich, L. & Burns, K.C., 'Body size determines rates of seed dispersal by giant king crickets', *Population Ecology 53*, 2011, pp. 73–80

Koenen, E.J.M., Clarkson, J.J., Pennington, T.D. & Chatrou, L.W., 'Recently evolved diversity and convergent radiations of rainforest mahoganies (Meliaceae) shed new light on the origins of rainforest hyperdiversity', *New Phytologist 207*, 2015, pp. 327–39

Kuschel, G., 'Beetles in a suburban environment: A New Zealand case study. The identity and status of Coleoptera in the natural and modified habitats of Lynfield, Auckland (1974–1989)', *DSIR Plant Protection Report 3*, 1990, pp. 1–119

Larivière, M.-C., 'Cixiidae (Insecta: Hemiptera, Auchenorrhyncha)', *Fauna of New Zealand 40*, 1999, pp. 1–93

Larivière, M.-C., Fletcher, M.J. & Larochelle, A., 'Auchenorrhyncha (Insecta: Hemiptera): catalogue', *Fauna of New Zealand 63*, 2010, pp. 1–232

Larivière, M.-C. & Larochelle, A., 'An overview of flat bug genera (Hemiptera, Aradidae) from New Zealand, with considerations on faunal diversification and affinities', *Denisia 19*, 2006, pp. 181–214

Lee, D.E., Bannister, J.M. & Ferguson, D.K., 'Taphonomy of leaf assemblages from a late Oligocene lignite, and an early Miocene diatomite deposit, southern New Zealand', in *Abstract Volume, 12th International Palynological Congress IPC-XII & 8th International Organisation of Palaeobotany Conference IOPC-VIII. Terra Nostra 2008/2*, Organizing Committee in Bonn (ed.) (Bonn: Schriften der GeoUnion Alfred-Wegener-Stiftung, Germany, 2008), pp. 139–40

Lee, D.E., Bannister, J.M., Kaulfuss, U., Conran, J.G. & Mildenhall, D.C., 'A fossil *Fuchsia* (Onagraceae) flower and an anther mass with *in situ* pollen from the early Miocene of New Zealand', *American Journal of Botany 100*, 2013, pp. 2052–65

Lee, D.E., Bannister, J.M., Raine, J.I. & Conran, J.G., 'Euphorbiaceae: Acalyphoideae fossils from early Miocene New Zealand: *Mallotus–Macaranga* leaves, fruits, and inflorescence with *in situ Nyssapollenites endobalteus* pollen', *Review of Palaeobotany and Palynology 163*, 2010, pp. 127–38

Lee, D.E., Conran, J.G., Lindqvist, J.K., Bannister, J.M. & Mildenhall, D.C., 'New Zealand Eocene, Oligocene and Miocene macrofossil and pollen records and modern plant distributions in the Southern Hemisphere', *Botanical Review 78*, 2012, pp. 235–60

Lee, D.E., Kaulfuss, U., Bannister, J.M. & Conran, J.G., 'Biodiversity and palaeoecology of Hindon and Foulden Maars: Two early Miocene *Konservat-Lagerstätten* from New Zealand', in *Geological Society of Australia abstracts number 117, Palaeo Down Under 2, Adelaide 11–15 Jul 2016*, J.R. Laurie, P.D. Kruse, D.C. Garcia-Bellido & J.D. Holmes (eds) (Geological Society of Australia Inc.: 2016) p. 40

Lee, D.E., Kaulfuss, U., Conran, J.G., Bannister, J.M. & Lindqvist, J.K., 'Biodiversity and palaeoecology of Foulden Maar: An early Miocene *Konservat-Lagerstätte* deposit in southern New Zealand', *Alcheringa 40*, 2016, pp. 525–41

Lee, D.E., Kaulfuss, U. & Lindqvist, J.K., 'An overview of the fauna of Foulden Maar – terrestrial life in New Zealand at the Oligocene–Miocene boundary', in *Geosciences 2011. Geoscience Society of New Zealand annual conference, 27 Nov–1 Dec 2011, GSNZ Miscellaneous Publications 130A*, N.J. Litchfield & K. Clark (eds) (Nelson: GSNZ, 2011) pp. 6

Lee, D.E., Lee, W.G., Jordan, G.J. & Barreda, V.D., 'The Cenozoic history of New Zealand temperate rainforests: Comparisons with southern Australia and South America', *New Zealand Journal of Botany 54*, 2016, pp. 100–27

Lee, D.E., Lee, W.G. & Mortimer, N., 'Where and why have all the flowers gone? Depletion and turnover in the New Zealand Cenozoic angiosperm flora in relation to palaeogeography and climate', *Australian Journal of Botany 49*, 2001, pp. 341–56

Lee, D.E., Lindqvist, J.K., Bannister, J.M., Raine, J.I., McDowall, R.M. & Harris, A.C., 'Exceptionally well preserved fossil flowers, fruit, fungi, leaves, sponges, fish and insects from an early Miocene forest-lake ecosystem: Foulden Maar, South Island, New Zealand', in *4th International Limnogeological Congress, 11–14 Jul 2007 (ILIC2007) – Limnogeology: Tales of an evolving Earth. Programme and abstract book*, ILIC (ed.) (Cosmocaixa, Barcelona: ILIC, 2007), p. 183

Lee, D.E., McDowall, R.M. & Lindqvist, J.K., '*Galaxias* fossils from Miocene lake deposits, Otago, New Zealand: The earliest records of the Southern Hemisphere family Galaxiidae (Teleostei)', *Journal of the Royal Society of New Zealand 37*, 2007, pp. 109–30

Lens, F., Jansen, S., Caris, P., Serlet, L. & Smets, E., 'Comparative wood anatomy of the Primuloid Clade (Ericales s.l)', *Systematic Botany 30*, 2005, pp. 163–83

Lenz, O.K. & Wilde, V., 'Changes in Eocene plant diversity and composition of vegetation: The lacustrine archive of Messel (Germany)', *Paleobiology 44*, 2018, pp. 709–35

Lenz, O.K., Wilde, V. & Riegel, W., 'Short-term fluctuations in vegetation and phytoplankton during the Middle Eocene greenhouse climate: A 640-kyr record from the Messel oil shale (Germany)', *International Journal of Earth Sciences 100*, 2011, pp. 1851–74

Lenz, O.K., Wilde, V. & Riegel, W., 'ENSO- and solar-driven sub-Milankovitch cyclicity in the Palaeogene greenhouse world; high-resolution pollen records from Eocene Lake Messel, Germany', *Journal of the Geological Society 174*, 2016, pp. 110–28

Leschen, R.A.B., Lawrence, J.F., Kuschel, G., Thorpe, S. & Wang, Q., 'Coleoptera genera of New Zealand', *New Zealand Entomologist 26*, 2003, pp. 15–28

Lindqvist, J.K. & Lee, D.E., 'High-frequency paleoclimate signals from Foulden Maar, Waipiata Volcanic Field, southern New Zealand: An early Miocene varved lacustrine diatomite deposit', *Sedimentary Geology 222*, 2009, pp. 98–110

Little, P., 'Turning our taonga into pet food', *New Zealand Herald*, 19 May 2019: www.nzherald.co.nz/nz/paul-little-so-much-to-learn-about-fossil-treasure-trove/SSARBFD3LMWUMIKPPUE2W4X4QQ

Lorenz, V., 'Maar-diatreme volcanoes, their formation, and their setting in hard-rock or soft-rock environments', *GeoLines 15*, 2003, pp. 72–83

Liu, G. & Leopold, E.B., 'Paleoecology of a Miocene Flora from the Shanwang Formation, Shandong Province, Northern

East China', *Palynology 16*, 1992, pp. 187–212

Lüthi, D., Le Floch, M., Bereiter, B., Blunier, T., Barnola, J.-M., Siegenthaler, U., Raynaud, D., Jouzel, J., Fischer, H., Kawamura, K. & Stocker, T.F., 'High-resolution carbon dioxide concentration record 650,000–800,000 years before present', *Nature 453*, 2008, pp. 379–82

Lutz, H. & Kaulfuss, U., 'A dynamic model for the meromictic lake Eckfeld Maar (Middle Eocene, Germany)', *Zeitschrift der Deutschen Gesellschaft für Geowissenschaften 157*, 2006, pp. 433–50

Lutz, H., Kaulfuss, U., Wappler, T., Löhnertz, W., Wilde, V., Mertz, D.F., Mingram, J., Franzen, J.L., Frankenhäuser, H. & Koziol, M., 'Eckfeld Maar: Window into an Eocene terrestrial habitat in Central Europe', *Acta Geologica Sinica 84*, 2010, pp. 984–1009

Mabberley, D.J., 'Meliaceae', in *The Families and Genera of Vascular Plants vol. 10. Flowering Plants. Eudicots: Sapindales, Cucurbitales, Myrtaceae*, K. Kubitzki (ed.) (Berlin: Springer Verlag, 2011), pp. 185–211

Mabberley, D.J., *Mabberley's Plant-book: A portable dictionary of plants, their classification and uses*, 4th edn (Cambridge: Cambridge University Press, 2017)

Macfarlane, R.P., Maddison, P.A., Andrew, I.G., Berry, J.A., Johns, P.M., Hoare, R.J.B., Larivière, M.-C., Greenslade, P., Henderson, R.C., Smithers, C.N., Palma, R.L., Ward, J.B., Pilgrim, R.L.C., Towns, D.R., McLellan, I., Teulon, D.A.J., Hitchings, T.R., Eastop, V.F., Martin, N.A., Fletcher, M.J., Stufkens, M.A.W., Dale, P.J., Burckhardt, D., Buckley, T.R. & Trewick, S.A., 'Phylum Arthropoda subphylum Hexapoda: Protura, springtails, Diplura, and insects', in *New Zealand Inventory of Biodiversity vol. 2: Kingdom Animalia: Chaetognatha, Ecdysozoa, ichnofossils*, D.P. Gordon (ed.) (Christchurch: Canterbury University Press, 2011), pp. 233–467

Maciunas, E., Conran, J.G., Bannister, J.M., Paull, R. & Lee, D.E., 'Miocene *Astelia* (Asparagales: Asteliaceae) macrofossils from southern New Zealand', *Australian Systematic Botany 24*, 2011, pp. 19–31

Macmillan, B.W., 'Biological flora of New Zealand. 7. *Ripogonum scandens* J.R. et G. Forst. (Smilacaceae), Supplejack, Kareao', *New Zealand Journal of Botany 10*, 1972, pp. 641–72

MacPhail, M.K., 'Fossil and modern *Beilschmiedia* (Lauraceae) pollen in New Zealand', *New Zealand Journal of Botany 18*, 1980, pp. 453–57

Manconi, R. & Pronzato, R., 'Global diversity of sponges (Porifera: Spongillina) in freshwater', *Hydrobiologia 595*, 2008, pp. 27–33

Massalongo, A.B., 'Vorläufige Nachricht über die neueren paläontologischen Entdeckungen am Monte Bolca',

Neues Jahrbuch für Mineralogie, Geognosie, Geologie und Petrefakten-kunde 1857, 1857, pp. 775–78

McDowall, R.M., *The Reed Field Guide to New Zealand Freshwater Fishes* (Auckland: Reed Publishing, 2000)

McPhee, E., 'Plaman withdraws; but maar still vulnerable', *ODT*, 1 Aug 2019: www.odt.co.nz/news/dunedin/plaman-withdraws-maar-still-vulnerable

Mehl, J., '*Eorchis miocaenica* nov.gen., nov.sp. aus dem Ober-Miozän von Öhningen, der bisher älteste fossile Orchideen-Fund', *Berichte aus den Arbeitskreisen Heimische Orchideen (Hanau) 12*, 1984, pp. 9–21

Meylan, B.A. & Butterfield, B.G., 'The structure of New Zealand woods', *New Zealand Department of Science and Industry Research Bulletin 222*, 1978, pp. 1–250

Mildenhall, D.C., Kennedy, E.M., Lee, D.E., Kaulfuss, U., Bannister, J.M., Fox, B. & Conran, J.G., 'Palynology of the early Miocene Foulden Maar, Otago, New Zealand: Diversity following destruction', *Review of Palaeobotany and Palynology 204*, 2014, pp. 27–42

Mildenhall, D.C., Kennedy, E.M., Prebble, J.G. & Shepherd, C.L., 'A distinctive diporate pollen grain (Apocynaceae?) from the Late Oligocene–Early Miocene of New Zealand', *New Zealand Journal of Geology and Geophysics 57*, 2014, pp. 264–68

Miller, T., 'Council now formally opposed to expansion of mining project', *ODT*, 14 Jun 2019: www.odt.co.nz/news/dunedin/council-now-formally-opposed-expansion-mine

Morgan, P.G., *Minerals and Mineral Substances of New Zealand* (New Zealand Geological Survey Bulletin 32) (Wellington: Government Printer, 1927)

Morris, C., 'Bridge nearing completion', *ODT Times*, 24 Jun 2019: www.odt.co.nz/news/dunedin/bridge-nearing-completion

Morris, C., 'DCC starts Foulden Maar process', *ODT*, 5 Nov 2019: www.odt.co.nz/news/dunedin/dcc/dcc-starts-foulden-maar-process

Morse, J.C., Frandsen, P.B., Graf, W. & Thomas, J.A., 'Diversity and ecosystem services of Trichoptera', *Insects 10*, 2019, p. 125

Mortimer, N. & Campbell, H.J., *Zealandia: Our continent revealed* (Auckland: Penguin, 2014)

Mortimer, N., Campbell, H.J., Tulloch, A.J., King, P.R., Stagpoole, V.M., Wood, R.A., Rattenbury, M.S., Sutherland, R., Adams, C.J., Collot, J. & Seton, M., 'Zealandia: Earth's hidden continent', *GSA Today 27*, 2017, pp. 27–35

Norton, S.A., 'Thrips pollination in the lowland forest of New Zealand', *New Zealand Journal of Ecology 7*, 1984, pp. 157–64

Nucete, M., van Konijnenburg-van Cittert, J.H.A. & van Welzen, P.C., 'Fossils and palaeontological distributions of

Macaranga and *Mallotus* (Euphorbiaceae)', *Palaeogeography, Palaeoclimatology, Palaeoecology 353–355*, 2012, pp. 104–15

Nudds, J.R. & Selden, P., *Extraordinary Fossil Ecosystems of North America* (London :Manson Publishing, 2008)

NZ Native Orchid Group, *New Zealand Native Orchids*, 2021: www.nativeorchids.co.nz/

Oliver, W.R.B., 'The flora of the Waipaoa Series (later Pliocene) of New Zealand', *Transactions of the New Zealand Institute 59*, 1928, pp. 287–303

Oliver, W.R.B., 'The Tertiary flora of the Kaikorai Valley, Otago, New Zealand', *Transactions of the Royal Society of New Zealand 66*, 1936, pp. 284–304

Oram, R., 'Foulden Maar: Unjustifiable vandalism and grand promises', *RadioNZ*, 24 May 2019: www.newsroom.co.nz/rod-orams-foulden-maar

Perrie, L.R., Field, A.R., Ohlsen, D.J. & Brownsey, P.J., 'Expansion of the fern genus *Lecanopteris* to encompass some species previously included in *Microsorum* and *Colysis* (Polypodiaceae)', *Blumea 66*, 2021, pp. 242–48

Pillon, Y., Lucas, E., Johansen, J.B., Sakishima, T., Hall, B., Geib, S.M. & Stacy, E.A., 'An expanded *Metrosideros* (Myrtaceae) to include *Carpolepis* and *Tepualia* based on nuclear genes', *Systematic Botany 40*, 2015, pp. 782–90

Poinar, G.O., Jr., 'Beetles with orchid pollinaria in Dominican and Mexican amber', *American Entomologist 62*, 2016a, pp. 172–77

Poinar, G.O., Jr., 'Orchid pollinaria (Orchidaceae) attached to stingless bees (Hymenoptera: Apidae) in Dominican amber', *Neues Jahrbuch für Geologie und Paläontologie, Abhandlungen 279*, 2016b, pp. 287–93

Poinar, G.O., Jr. & Rasmussen, F.N., 'Orchids from the past, with a new species in Baltic amber', *Botanical Journal of the Linnean Society 183*, 2017, pp. 327–33

Pole, M., 'Early Miocene flora of the Manuherikia Group, New Zealand. 1. Ferns', *Journal of the Royal Society of New Zealand 22*, 1992a, pp. 279–86

Pole, M., 'Early Miocene flora of the Manuherikia Group, New Zealand. 2. Conifers', *Journal of the Royal Society of New Zealand 22*, 1992b, pp. 287–302

Pole, M., 'Early Miocene flora of the Manuherikia Group, New Zealand. 5. Smilacaceae, Polygonaceae, Elaeocarpaceae', *Journal of the Royal Society of New Zealand 23*, 1993a, pp. 289–302

Pole, M., 'Early Miocene flora of the Manuherikia Group, New Zealand. 7. Myrtaceae, including *Eucalyptus*', *Journal of the Royal Society of New Zealand 23*, 1993b, pp. 313–28

Pole, M., 'Early Miocene flora of the Manuherikia Group, New Zealand. 8. *Nothofagus*', *Journal of the Royal Society of New Zealand 23*, 1993c, pp. 329–44

Pole, M., 'Miocene broad-leaved *Podocarpus* from Foulden Hills, New Zealand', *Alcheringa 17*, 1993d, pp. 173–77

Pole, M., '*Nothofagus* from the Dunedin Volcanic Group (Mid–Late Miocene), New Zealand', *Alcheringa 17*, 1993e, pp. 77–90

Pole, M., 'Deciduous *Nothofagus* leaves from the Miocene of Cornish Head, New Zealand', *Alcheringa 18*, 1994a, pp. 79–83

Pole, M., 'The New Zealand flora – entirely long-distance dispersal?', *Journal of Biogeography 21*, 1994b, pp. 625–35

Pole, M., 'Plant macrofossils from the Foulden Hills Diatomite (Miocene), Central Otago, New Zealand', *Journal of the Royal Society of New Zealand 26*, 1996, pp. 1–39

Pole, M., 'Miocene conifers from the Manuherikia Group, New Zealand', *Journal of the Royal Society of New Zealand 27*, 1997, pp. 355–70

Pole, M., 'The Proteaceae record in New Zealand', *Australian Systematic Botany 11*, 1998, pp. 343–72

Pole, M., 'Conifer and cycad distribution in the Miocene of southern New Zealand', *Australian Journal of Botany 55*, 2007a, pp. 143–64

Pole, M., 'Monocot macrofossils from the Miocene of southern New Zealand', *Palaeontologia Electronica 10*, 2007b, pp. 3.15A, 1–21

Pole, M., 'Dispersed leaf cuticle from the early Miocene of southern New Zealand', *Palaeontologia Electronica 11*, 2008, pp. 15A, 1–117

Pole, M., 'Plant macrofossils', in *New Zealand Inventory of Biodiversity vol. 3: Kingdoms Bacteria, Protozoa, Chromista, Plantae, Fungi*, D.P. Gordon (ed.) (Christchurch: Canterbury University Press, 2012), pp. 460–75

Pole, M., Dawson, J. & Denton, T., 'Fossil Myrtaceae from the early Miocene of southern New Zealand', *Australian Journal of Botany 56*, 2008, pp. 67–81

Pole, M., Douglas, B. & Mason, G., 'The terrestrial Miocene biota of southern New Zealand', *Journal of the Royal Society of New Zealand 33*, 2003, pp. 415–26

Pole, M. & Moore, P.R., 'A late Miocene leaf assemblage from Coromandel Peninsula, New Zealand, and its climatic implications', *Alcheringa 35*, 2010, pp. 103–21

Poschmann, M., Schindler, T. & Uhl, D., 'Fossil-*Lagerstätte* Enspel – a short review of current knowledge, the fossil association, and a bibliography', *Palaeobiodiversity and Palaeoenvironments 90*, 2010, pp. 3–20

Raine, J.I., Beu, A.G., Boyes, A.F., Campbell, H.J., Cooper, R.A., Crampton, J.S., Crundwell, M.P., Hollis, C.J., Morgans, H.E.G. & Mortimer, N., 'New Zealand Geological Timescale NZGT 2015/1', *New Zealand Journal of Geology and Geophysics 58*, 2015, pp. 398–403

Raine, J.I., Mildenhall, D.C. & Kennedy, E.M., *New Zealand Fossil Spores and Pollen: An illustrated catalogue*, 4th edn (GNS Science Miscellaneous Series No. 4), 2011: www.gns. cri.nz/what/earthhist/fossils/spore_pollen/catalog/index. htm, 853 html pages; issued also in CD version

Ramírez, S.R., Gravendeel, B., Singer, R.B., Marshall, C.R. & Pierce, N.E., 'Dating the origin of the Orchidaceae from a fossil orchid with its pollinator', *Nature 448*, 2007, pp. 1042–45

Rasser, M.W., Bechly, G., Böttcher, R., Ebner, M., Heizmann, E.P.J., Höltke, O., Joachim, C., Kern, A.K., Kovar-Eder, J., Nebelsick, J.H., Roth-Nebelsick, A., Schoch, R.R., Schweigert, G. & Ziegler, R., 'The Randeck Maar: Palaeoenvironment and habitat differentiation of a Miocene lacustrine system', *Palaeogeography, Palaeoclimatology, Palaeoecology 392*, 2013, pp. 426–53

Raymo, M.E., Lisiecki, L.E. & Nisancioglu, K.H., 'Plio-Pleistocene ice volume, Antarctic climate, and the global $\delta^{18}O$ record', *Science 313*, 2006, pp. 492–95

Recher, H.F., Lunney, D. & Dunn, I. (eds), *A Natural Legacy: Ecology in Australia*, 2nd edn (Sydney: Pergamon Press, 1986)

Reichgelt, T., 'Reconstructing southern New Zealand Miocene terrestrial climate and ecosystems from plant fossils', PhD thesis, University of Otago, Dunedin, 2015

Reichgelt, T., D'Andrea, W.J. & Fox, B.R.S., 'Abrupt plant physiological changes in southern New Zealand at the termination of the Mi-1 event reflect shifts in hydroclimate and pCO$_2$', *Earth and Planetary Science Letters 455*, 2016, pp. 115–24

Reichgelt, T., D'Andrea, W.J., Valdivia-McCarthy, A.C., Fox, B.R.S., Bannister, J.M., Conran, J.G., Lee, W.G. & Lee, D.E., 'Elevated CO$_2$, increased leaf-level productivity, and water-use efficiency during the early Miocene', *Climate of the Past 16*, 2020, pp. 1509–21

Reichgelt, T., Jones, W.A., Jones, D.T., Conran, J.G., Bannister, J.M., Kennedy, E.M., Mildenhall, D.C. & Lee, D.E., 'The flora of Double Hill (Dunedin Volcanic Complex, middle–late Miocene)', *Journal of the Royal Society of New Zealand 44*, 2014, pp. 105–35

Reichgelt, T., Kennedy, E.M., Conran, J.G., Lee, W.G. & Lee, D.E., 'The presence of moisture deficits in Miocene New Zealand', *Global and Planetary Change 172*, 2019, pp. 268–77

Reichgelt, T., Kennedy, E.M., Mildenhall, D.C., Conran, J.G., Greenwood, D.R. & Lee, D.E., 'Quantitative palaeoclimate estimates for early Miocene southern New Zealand: Evidence from Foulden Maar', *Palaeogeography, Palaeoclimatology, Palaeoecology 378*, 2013, pp. 36–44

Renner, S.S., Strijk, J.S., Strasberg, D. & Thébaud, C., 'Biogeography of the Monimiaceae (Laurales): A role for East Gondwana and long-distance dispersal, but not West Gondwana', *Journal of Biogeography 37*, 2010, pp. 1227–38

Romero, E.J. & Hickey, L.J., 'A fossil leaf of Akaniaceae from Paleocene beds in Argentina', *Bulletin of the Torrey Botanical Club 103*, 1976, pp. 126–31

Rozefelds, A.C. & Christophel, D.C., 'Cenozoic *Elaeocarpus* (Elaeocarpaceae) fruits from Australia', *Alcheringa 26*, 2002, pp. 261–74

Ruiz, A.I., Guantay, M.E. & Ponessa, G.I., 'Leaf morphology, anatomy and foliar architecture of *Myrsine laetevirens* (Myrsinaceae)', *Lilloa 49*, 2012, pp. 59–67

Sauquet, H., Weston, P.H., Anderson, C.L., Barker, N.P., Cantrill, D.J., Mast, A.R. & Savolainen, V., 'Contrasted patterns of hyperdiversification in Mediterranean hotspots', *Proceedings of the National Academy of Sciences 106*, 2009, pp. 221–25

Schaal, S. & Ziegler, W. (eds), *Messel: An insight into the history of life and of the earth* (Oxford: Oxford University Press, 1992)

Schindler, T. & Wuttke, M., 'Geology and limnology of the Enspel Formation (Chattian, Oligocene; Westerwald, Germany)', *Palaeobiodiversity and Palaeoenvironments 90*, 2010, pp. 21–27

Schmidt, A.R., Kaulfuss, U., Bannister, J.M., Baranov, V., Beimforde, C., Bleile, N., Borkent, A., Busch, A., Conran, J.G., Engel, M.S., Harvey, M., Kennedy, E.M., Kerr, P.H., Kettunen, E., Kiecksee, A.P., Lengeling, F., Lindqvist, J.K., Maraun, M., Mildenhall, D.C., Perrichot, V., Rikkinen, J., Sadowski, E.-M., Seyfullah, L.J., Stebner, F., Szwedo, J., Ulbrich, P. & Lee, D.E., 'Amber inclusions from New Zealand', *Gondwana Research 56*, 2018, pp. 135–46

Schulz, R., Bunes, H., Babriel, G., Pucher, R., Rolf, C., Wiederhold, H. & Wonik, T., 'Detailed investigation of preserved maar structures by combined geophysical surveys', *Bulletin of Volcanology 68*, 2005, pp. 95–106

Schwarzhans, W., Scofield, R.P., Tennyson, A.J.D., Worthy, J.P. & Worthy, T.H., 'Fish remains, mostly otoliths, from the non-marine early Miocene of Otago, New Zealand', *Acta Palaeontologica Polonica 57*, 2011, pp. 319–50

Scott, J.M., Pontesilli, A., Brenna, M., White, J.D.L., Giacalone, E., Palin, J.M. & le Roux, P.J., 'The Dunedin Volcanic Group and a revised model for Zealandia's alkaline intraplate volcanism', *New Zealand Journal of Geology and Geophysics 63*, 2020, pp. 510–29

Selden, P.A. & Kaulfuss, U., 'Fossil arachnids from the earliest Miocene Foulden Maar Fossil-Lagerstätte, New Zealand', *Alcheringa 43*, 2018, pp. 165–69

Selden, P. & Nudds, J., *Fossil Ecosystems of North America: A guide to the sites and their extraordinary biotas* (Boca Raton, Florida: CRC Press, 2008)

Shepherd, L.D., de Lange, P.J., Townsend, A. & Perrie, L.R., 'A biological and ecological review of the endemic New Zealand genus *Alseuosmia* (toropapa; Alseuosmiaceae)', *New Zealand Journal of Botany 58*, 2020, pp. 2–18

Shirley, C., *Forest Fungi Quick Reference Guide*, version 5.01 (published online by author, 2020): http://hiddenforest.co.nz/fungi/

Sirvid, P.J., Zhang, Z.-Q., Harvey, M.S., Rhode, B.E., Cook, D.R., Bartsch, I. & Staples, D.A., 'Phylum Arthropoda Chelicerata: horseshoe crabs, arachnids, sea spiders', in *New Zealand Inventory of Biodiversity. Volume 2. Kingdom Animalia. Chaetognatha, Ecdysozoa, ichnofossils*, D.P. Gordon (ed.) (Christchurch: Canterbury University Press, 2011), pp. 50–89

Steart, D.C., Greenwood, D.R. & Boon, P.I., 'The chemical constraints upon leaf decay rates: Taphonomic implications among leaf species in Australian terrestrial and aquatic environments', *Review of Palaeobotany and Palynology 157*, 2009, pp. 358–74

Stedman, M., 'Opinion Piece – Clarity hard to find in shadows', *ODT*, 24 May 2019: www.odt.co.nz/opinion/clarity-hard-find-shadows

Steinthorsdottir, M., Vajda, V. & Pole, M., 'Significant transient pCO_2 perturbation at the New Zealand Oligocene–Miocene transition recorded by fossil plant stomata', *Palaeogeography, Palaeoclimatology, Palaeoecology 515*, 2019, pp. 152–61

Strogen, D.P., Higgs, K.E., Griffin, A.G. & Morgans, H.E.G., 'Late Eocene–early Miocene facies and stratigraphic development, Taranaki Basin, New Zealand: The transition to plate boundary tectonics during regional transgression', *Geological Magazine 156*, 2019, pp. 1751–70

Sun, M., Naeem, R., Su, J.-X., Cao, Z.-Y., Burleigh, J.G., Soltis, P.S., Soltis, D.E. & Chen, Z.-D., 'Phylogeny of the Rosidae: A dense taxon sampling analysis', *Journal of Systematics and Evolution 54*, 2016, pp. 363–91

Sutherland, J.I., 'Miocene petrified wood and associated borings and termite faecal pellets from Hukatere Peninsula, Kaipara Harbour, North Auckland, New Zealand', *Journal of the Royal Society of New Zealand 33*, 2003, pp. 395–414

Tennyson, A.J.D., Scofield, R.P. & Worthy, T.H., 'Vertebrate survivors of Zealandia's Oligocene "drowning"', in *Abstract volume, Geosciences 2011 Conference, Nelson, New Zealand*. Geoscience Society of New Zealand Miscellaneous Publication 130A, N.J. Litchfield & K. Clark (eds) (Nelson: Geoscience Society of New Zealand, 2011), p. 107

Testo, W.L., Field, A.R., Sessa E.B. & Sundue, M., 'Phylogenetic and morphological analyses support the resurrection of *Dendroconche* and the recognition of two new genera in Polypodiaceae subfamily Microsoroideae', *Systematic Botany 44*, 2019, pp. 737–52

Thompson, B., Brathwaite, R.L. & Christie, A.B., *Mineral Wealth of New Zealand*, IGNS information series 33 (Lower Hutt: Institute of Geological & Nuclear Sciences, 1995)

Thorsen, M.J., Dickinson, K.J.M. & Seddon, P.J., 'Seed dispersal systems in the New Zealand flora', *Perspectives in Plant Ecology, Evolution and Systematics 11*, 2009, pp. 285–309

Travis, C., 'Geology of the Slip Hill area east of Middlemarch', MSc thesis, University of Otago, Dunedin, 1965

Trewick, S.A., Paterson, A.M. & Campbell, H.J., 'Hello New Zealand', *Journal of Biogeography 34*, 2007, pp. 1–6

Tsutsumi, C. & Kato, M., 'Evolution of epiphytes in Davalliaceae and related ferns', *Botanical Journal of the Linnean Society 151*, 2006, pp. 495–510

US Geological Survey, *Mineral Commodity Summaries, January 2021* (US Geological Survey, Reston, VA, 2021): https://pubs.usgs.gov/periodicals/mcs2021/mcs2021-diatomite.pdf

Vanner, M.R., Conran, J.G., Bannister, J.M. & Lee, D.E., 'Cenozoic conifer wood from the Gore Lignite Measures, Southland, New Zealand', *New Zealand Journal of Botany 56*, 2018, pp. 291–310

von Konrat, M.J., Braggins, J.E. & de Lange, P.J., '*Davallia* (Pteridophyta) in New Zealand, including description of a new subspecies of *D. tasmanii*', *New Zealand Journal of Botany 37*, 1999, pp. 579–93

Wallis, G.P. & Jorge, F., 'Going under down under? Lineage ages argue for extensive survival of the Oligocene marine transgression on Zealandia', *Molecular Ecology 27*, 2018, pp. 4368–96

Wardle, P. & MacRae, A.H., 'Biological flora of New Zealand. 1. *Weinmannia racemosa*', *New Zealand Journal of Botany 4*, 1966, pp. 114–31

Waters, S.B., 'A Cretaceous dance fly (Diptera: Empididae) from Botswana', *Systematic Entomology 14*, 1989, pp. 233–41

Watters, W.A., 'Ulrich, George Henry Frederick', in *Te Ara – the Encyclopedia of New Zealand*: https://teara.govt.nz/en/biographies/2u1/ulrich-george-henry-frederick

Webb, L.J., 'A physiognomic classification of Australian rain forests', *Journal of Ecology 47*, 1959, pp. 551–70

White, J.D.L. & Ross, P.S., 'Maar-diatreme volcanoes: A review', *Journal of Volcanology and Geothermal Research 201*, 2011, pp. 1–29

Wilde, V. & Frankenhäuser, H., 'The middle Eocene plant taphocoenosis from Eckfeld (Eifel, Germany)', *Review of Palaeobotany and Palynology 101*, 1998, pp. 7–28

Wilson, G.S., Pekar, S.F., Naish, T.R., Passchier, S. & DeConto, R., 'The Oligocene–Miocene boundary – Antarctic climate response to orbital forcing', in *Developments in Earth and Environmental Sciences vol. 8*, F. Florindo & M. Siegert (eds) (Amsterdam: Elsevier, 2008), pp. 369–400

Wolfe, J.A., 'A method of obtaining climatic parameters from leaf assemblages', *Bulletin of the United States Geological Survey 2040*, 1993, pp. 1–71

Worthy, T.H., Hand, S.J., Worthy, J.P., Tennyson, A.J.D. & Scofield, P.R., 'A large fruit pigeon (Columbidae) from the early Miocene of New Zealand', *The Auk 126*, 2009, pp. 649–56

Worthy, T.H., Tennyson, A.J.D., Jones, C., McNamara, J.A. & Douglas, B.J., 'Miocene waterfowl and other birds from central Otago, New Zealand', *Journal of Systematic Palaeontology 5*, 2007, pp. 1–39

Worthy, T.H., Tennyson, A.J.D. & Scofield, R.P., 'An early Miocene diversity of parrots (Aves, Strigopidae, Nestorinae) from New Zealand', *Journal of Vertebrate Paleontology 31*, 2011, pp. 1102–16

Wuttke, M., 'Conservation – dissolution – transformation. On the behaviour of biogenic materials during fossilization', in *Messel. An insight into the history of life and of the earth*, S. Schaal & W. Ziegler (eds) (Oxford, UK: Oxford University Press, 1992), pp. 265–75

Wuttke, M., Uhl, D. & Schindler, T., 'The Fossil-Lagerstätte Enspel – exceptional preservation in an Upper Oligocene maar', *Palaeobiodiversity and Palaeoenvironments 90*, 2010, pp. 1–2

Wuttke, M., Schindler, T. & Smith, K.T., 'The Fossil-Lagerstätte Enspel – a crater lake in a volcanic-influenced terrestrial environment of the Westerwald Basin (late Oligocene, Western Germany)', *Palaeobiodiversity and Palaeoenvironments 95*, 2015, pp. 1–4

Yamasaki, E. & Sakai, S., 'Wind and insect pollination (ambophily) of *Mallotus* spp. (Euphorbiaceae) in tropical and temperate forests', *Australian Journal of Botany 61*, 2013, pp. 60–66

Yang, H. & Yang, S., 'The Shanwang fossil biota in eastern China: A Miocene *Konservat-Lagerstätte* in lacustrine deposits', *Lethaia 27*, 1994, pp. 345–54

Youngson, J.H., 'Mineralized vein systems and Miocene maar crater sediments at Hindon, East Otago, New Zealand', MSc thesis, University of Otago, Dunedin, 1993

Zachos, J.C., Dickens, G.R. & Zeebe, R.E., 'An early Cenozoic perspective on greenhouse warming and carbon-cycle dynamics', *Nature 451*, 2008, pp. 279–83

Zhang, Y.G., Pagani, M., Liu, Z., Bohaty, S.M. & DeConto, R., 'A 40-million-year history of atmospheric CO_2', *Philosophical Transactions of the Royal Society A: Mathematical, Physical and Engineering Sciences 371*, 2013, p. 20130096

ACKNOWLEDGEMENTS

We particularly wish to acknowledge the following researchers who committed time, energy and insights into revealing the fossil treasures of Foulden Maar. Without them, this book could not have been written.

A With a degree in botany from the University of Nottingham and having studied paleobotany at the University of London, **Jennifer Bannister** has been an integral part of the Foulden research team from the beginning. Her discovery of the first fossil flower led to the 2003 Harold Wellman Prize, awarded each year by the Geoscience Society of New Zealand for 'an outstanding discoverer of New Zealand fossils and his/her appreciation of the important role of fossil evidence in the resolution of New Zealand geology'.

B **Jon Lindqvist** has a PhD in sedimentology from the University of Otago. He coined the term 'Foulden Maar' and was the first to appreciate the significance of the climate record preserved in the fine laminations seen in the diatomite. Jon also developed a technique of cutting vertical columns of fresh/wet diatomite with a chainsaw, for correlation between the various mining pits and the cores and for investigation of the climate signals.

C **Tony Harris** (far left) with John Conran, Jon Lindqvist, the late Peter Bannister and Jennifer Bannister. Tony has a long career as an entomologist based at the Otago Museum and for decades has published short informative articles on all aspects of natural history in the *Otago Daily Times*. Tony was co-author of papers on the first insects to be collected at Foulden: the scale insects and an ant. He also has the distinction of formally describing the first fossil insect from New Zealand, a bibionid fly larva of Eocene age from North Otago.

D **Dallas Mildenhall** is a paleobotanist from GNS Science in Lower Hutt and has been integral to the research at Foulden. A graduate of Victoria University of Wellington, Dallas has been involved in the development and application of palynological and other techniques in research on paleoclimate, paleoenvironment, biostratigraphy, dating, evolution and plant taxonomy.

E **Liz Kennedy** obtained her MSc from the University of Canterbury and her PhD from the Open University, UK. Her primary area of research is in applied paleobotany, focusing on the application of plant fossil-based multi-proxy approaches to paleoclimate and paleoenvironmental reconstruction.

F **Ian Raine** was co-author on the paper to describe *Fouldenia staminosa*, the first fossil flower from Foulden Maar. He has worked as a palynologist at GNS for several decades and is a co-author, with Liz Kennedy and Dallas Mildenhall, of the comprehensive illustrated catalogue of New Zealand fossil spores and pollen, which has been invaluable in research on Foulden palynomorphs.

G **Bethany Fox** is a graduate of the University of Cambridge and the Open University, UK, and completed a PhD on climate change at the Oligocene/Miocene boundary at the University of Otago in 2012. Beth is now based at the University of Huddersfield.

H **Tammo Reichgelt** came to the University of Otago from Utrecht University in the Netherlands and completed a PhD on the Miocene terrestrial climate from plant fossils in 2014. He is now a lecturer at the University of Connecticut.

I **Gary Wilson** was joint principal investigator for the first successful Marsden Grant for research at Foulden Maar and oversaw the drilling programme in 2009.

J **Andrew Gorman** was leader of the geophysics field school at which students carried out the preliminary geophysical investigations of Foulden Maar. With Gary Wilson, he co-supervised Daniel Jones' MSc thesis, which established the subsurface structure of the maar.

This book is the result of almost two decades of research on all aspects of Foulden Maar and its fossil treasures. We also wish to thank the Gibson family, the McRae family and other landowners in the area, for allowing geology students and staff from the University of Otago to roam over their land, making maps, collecting rock and fossil samples and carrying out geophysical surveys. The McRae family kindly allowed us to carry out the crucial Marsden-funded drilling project in 2009. We are grateful to the mining companies, who provided access for our research, and we are particularly indebted to Dr Alan Walker, who initially facilitated access to the mining site and encouraged our research.

We are also indebted to the many colleagues and students who have been and continue to play an integral part in research on Foulden Maar. Of the literally hundreds of people to thank, we acknowledge in particular:

- The support of many of our colleagues and students in the departments of geology, botany and zoology at the University of Otago, and the School of Biological Sciences at the University of Adelaide
- Stephen Read, who patiently prepared many maps, edited many photographs and offered wise advice
- Liz Girvan, Otago Centre for Electron Microscopy, University of Otago, for assistance with scanning electron microscopy imaging of fossils
- Margaret Harper, for information on the diatoms at an early stage in our research and for helping solve the taxonomic puzzle of the correct name to use for Foulden Maar diatoms
- Donald MacFarlan, who helped locate obscure mining publications and discussed mining issues on many occasions
- Bill Lee, Christina Beimforde, Billy D'Andrea, Ewan Fordyce, James White, Jeffrey Robinson, Jenny Stein, Belinda Smith-Lyttle, Brent Pooley, Ray Marx, Damian Wall, John Williams, Adrien Dever, David and Wyn Jones, Sophie White, Liz Maciunas, George Gibbs, Isaac Kerr, Kate Evans, Hamish Campbell, Nick Mortimer, Rich Leschen, Ian Henderson, Don Jellyman and the late Bob McDowall, who provided support for our research in many different ways
- Daniel Jones, Dianne Nyhof, Mathew Vanner, Ian Geary, Marcus Richards, Henry Gard, Nichole Moerhuis and the many students who collected fantastic fossils on numerous occasions
- The members of the 'Save Foulden Maar' campaign, who gathered more than 11,000 signatures on a petition to halt the proposed mining. Of the many, we especially thank Andrea Bosshard, Shane Loader, Nic Rawlence, Peter Hayden, Alan Mark, Kimberley Collins, Peter Wimsett, Tamsin Braisher, Rosemary Clucas and Robin Thomas. We thank the Thompson family for their support, Laura Harry and James Cambridge for pro bono advice, and John Youngson and members of the Puketeraki rūnaka who supported saving this fossil treasure.

- Many *Otago Daily Times* reporters, especially Simon Hartley, whose article set in motion the campaign to save Foulden Maar
- Support from the late Dave Cull, former mayor, and Sue Bidrose, former CEO of the Dunedin City Council, who were willing to listen to scientists and the local community about the scientific importance of Foulden Maar. We had many helpful discussions with Karilyn Canton, Phillip Marshall, Aaron Hawkins, Rachel Elder and Kate Wilson.
- Funding from Marsden Grants from the Royal Society of New Zealand in 2008, 2011 and 2014 and several Otago Research Grants from the Division of Sciences, University of Otago, which together made possible almost two decades of research at Foulden Maar
- Funding from the German Research Foundation (grant 429296833) for continued research on fossil insects from Foulden Maar
- The Royal Society of New Zealand Charles Fleming Publishing Award

Many people have provided photographs, figures and information for this book. We particularly thank Paul Selden, Nick Butterfield (University of Cambridge), Peter Kamp (University of Waikato), Rod Morris, Thomas Buckley and Birgit Rhode (Landcare Research), Ian Henderson (Massey University), Martin Koziol (Maarmuseum Manderscheid), Viktor Baranov (University of Munich), Jacek Szwedo (University of Gdansk), Gildas Gâteblé (Institut Agronomique néo-Calédonien), Frank Zich (Australian Tropical Herbarium) Torsten Wappler, Hessisches Landesmuseum Darmstadt and Josh Jenkins Shaw (Natural History Museum of Denmark). Paula Peeters provided the reconstruction drawing of Foulden Maar.

We would also like to thank the many individuals and organisations who sent submissions to the Overseas Investment Office, wrote letters to newspapers, and organised and attended talks on Foulden Maar. We thank the many artists, especially

Vivienne Robertson, and poets who contributed paintings and poems to various exhibitions. In particular, we thank Jennifer Eccles and Bruce Hayward and other members of the Geoscience Society of NZ, the NZ Ecological Society, Forest & Bird, and ECO (the Environment and Conservation Organisations of NZ) for their support for saving Foulden Maar.

We trust that the efforts of all these people and organisations will result in new measures to protect such precious fossil sites for future generations. This book is for them.

The advice of publishers Rachel Scott and Sue Wootton and the expertise of editor Imogen Coxhead and designer Fiona Moffat of Otago University Press are gratefully acknowledged.

INDEX

Page numbers in **bold** refer to illustrations